Proteomics Technologies and Applications

Edited by Ibrokhim Y. Abdurakhmonov

Published in London, United Kingdom

IntechOpen

Supporting open minds since 2005

Proteomics Technologies and Applications
http://dx.doi.org/10.5772/intechopen.73445
Edited by Ibrokhim Y. Abdurakhmonov

Contributors
Hanaa Amer, Saranyoo Ponnikorn, Sasipim Thitivirachawat, Siripath Peter Kong, Chanawin Tanjasiri, Sumalee Tungpradabkul, Suradej Hongeng, Xianquan Zhan, Xiaohan Zhan, Xiaowei Wang, Radu Albulescu, Cristian V.A. Munteanu, Stefana Petrescu, Ioana V Militaru, Mirela Sarbu, Raluca Ica, Alina-Diana Zamfir, Cristiana Tanase, Alice Grigore, Adrian Albulescu, Andrei Jose Petrescu, Alan Doucette, Andrew Crowell, Domenica Scumaci, Giovanni Cuda

First published in London, United Kingdom, 2019 by IntechOpen
IntechOpen is the global imprint of INTECHOPEN LIMITED, registered in England and Wales, registration number: 11086078, The Shard, 25th floor, 32 London Bridge Street
London, SE19SG – United Kingdom
Printed in Croatia

British Library Cataloguing-in-Publication Data
A catalogue record for this book is available from the British Library

Additional hard and PDF copies can be obtained from orders@intechopen.com

Proteomics Technologies and Applications
Edited by Ibrokhim Y. Abdurakhmonov
p. cm.
Print ISBN 978-1-78984-610-2
Online ISBN 978-1-78984-611-9
eBook (PDF) ISBN 978-1-83962-221-2

For EU product safety concerns:
IN TECH d.o.o., Prolaz Marije Krucifikse Kozulić 3, 51000 Rijeka, Croatia, info@intechopen.com or visit our website at intechopen.com.

Meet the editor

Ibrokhim Y. Abdurakhmonov received his BS (1997) in Biotechnology from the National University, his MS in Plant Breeding (2001) from Texas A&M University, USA, and his PhD (2002) in Molecular Genetics, his Doctor of Science (2009) in Genetics, and his full professorship (2011) in Molecular Genetics and Molecular Biotechnology from the Academy of Sciences of Uzbekistan. He founded (2012) and is currently leading the Center of Genomics and Bioinformatics of Uzbekistan. He serves as an associate editor/editorial board member of several international and national journals on plant sciences. He received a Government award – 2010 chest badge "Sign of Uzbekistan", 2010 TWAS prize, and "ICAC Cotton Researcher of the Year 2013" for his outstanding contribution to cotton genomics and biotechnology. He was elected as a TWAS Fellow (2014) in Agricultural Science and as a co-chair/chair of the "Comparative Genomics and Bioinformatics" Workgroup (2015) of the International Cotton Genome Initiative. He was elected (2017) as a member and Vice-President of the Academy of Sciences of Uzbekistan. He was appointed (2017) as Minister of Innovative Development of Uzbekistan.

Contents

Preface

Proteins or peptides are very important molecules of living organisms and play many vital functions in cell physiology. The whole set of proteins and peptides produced by an organism is called the proteome. The large-scale scientific study and experimental analysis of proteomes is referred to as *proteomics*.

Proteomics is an interdisciplinary science influenced by advances in many scientific directions such as chemistry, physics, biology, biochemistry, biophysics, genomics, bioinformatics, etc. Specifically, the latest advances in genomics and bioinformatics have greatly contributed to proteomics research, and have helped to elucidate the functionality of proteomes in terms of molecular functions and interactions in a given cell or tissue.

Proteomics research is equipped with a variety of detection, separation, quantification, and localization methods. Such methods include simple procedures of detection and quantification through chemical reactions specific to amino acids and high-throughput modern mass spectrometry tools.

The results of proteomics research are very useful for understanding the possible cause of affected cell/tissue or disease physiology and can serve as potential new ways to treat disease not only in general cases but also at a personalized level. Pivotal application of proteomics uses specific protein biomarkers to diagnose disease. A number of techniques allow testing for proteins produced during a particular disease, which helps to diagnose the disease quickly or to make reliable prognoses of its future appearance. Proteomics can also be used to reveal complex plant–insect interactions that help to identify genetic mechanisms in plant defenses and adaptation to different biotic and abiotic stresses.

Proteomics Technologies and Application includes six chapters from the international research community, covering both aspects of proteome analysis and its applications. The book provides readers with the latest advances in proteomics research tools and technologies using proteomics data to understand, diagnose, and treat various diseases. Chapters discuss the basic tools of protein detection, separation, and purification to advanced 2D-gel and mass spectrometry based on new generation technologies, which are useful for developing biomarkers and understanding, diagnosing, and curing various diseases.

These six chapters cover limited research and review topics in proteomics research and applications; however, they are useful for scientists, students, and readers of life sciences.

I appreciate the valuable contributions from all authors and thank the IntechOpen book department and publishing process manager(s) for their support during compiling this book.

Ibrokhim Y. Abdurakhmonov
Center of Genomics and Bioinformatics,
Academy of Sciences of Uzbekistan,
Tashkent, Uzbekistan

Section 1

Proteomics Approaches

Chapter 1

2D Gel Electrophoresis to Address Biological Issues

Domenica Scumaci and Giovanni Cuda

Abstract

Two-dimensional (2D) gel electrophoresis is a high-resolution technique for the study of proteome. This chapter describes how it can be applied to characterize specific differences in the proteome profile of breast cancer cells following gene target interference. The proteome is the complete set of proteins encoded by a genome, and proteomic analysis consists in profiling the whole proteins expressed in a given cell, tissue, organ, or organism. Proteomic expression has the main purpose of qualitatively and quantitatively comparing proteins expressed under physiological and/or pathological conditions. Although it is not the unique approach used in modern proteomics, two-dimensional electrophoresis (2DE) is unrivaled allowing simultaneous separation of thousands of proteins and the detection of post-translational modification, not predictable through genome analysis. 2DE combines two physical principles to separate complex protein mixtures: the isoelectric point and the molecular weight. The result is a gel map in which each protein isoform present in the sample can be visualized as a spot, analyzed, quantified, and identified by mass spectrometry analysis. Here we outline features and advantages of the 2DE-based proteomic approach and we describe how 2DE meets biochemistry and molecular biology to address specific issues.

Keywords: 2D gel electrophoresis, proteomics, mass spectrometry, breast cancer, western blot, image analysis, silver staining, trypsin digestion

1. Introduction

Proteins are the effectors of almost all functions in living cells. The proteome is the complete set of proteins encoded by a genome [1, 2]. For a given organism or living cells, genome is a static entity, except if a peculiar mutation occurs. Proteome is dynamic and its composition differs under the influence of specific physiological and or pathological stimuli. Proteomics is the science that describes the proteome and provides identification and quantification of proteins, including co- and post-translational modifications, their localization, interactions, activities, and functions [3].

Resolution of thousands of proteins in a single experiment implies the employment of high-throughput analytical tools with high resolution, sensitivity, and reproducibility. Two-dimensional gel electrophoresis is a key separation technique for proteomic research, useful for qualitative and quantitative protein expression profiling. The technique has been initially applied for protein separation in 1975 by O′ Farrel [4]; it has been used for over three decades, and it is still the most widely utilized protein separation method for proteome or sub-proteome analysis. The main advantage of 2DE is the capability of simultaneously resolving complex protein mixture, up to 10,000, allowing their quantitation as well as the analysis of all post-translational modifications [5]. 2DE has been adopted for a variety of applications: from drug discovery to diagnostics, therapy, and biochemistry [6–10].

In the last decade, 2DE has been significantly improved and currently it is a high-resolution, reliable, and reproducible method [11–13]. 2DE coupled together isoelectric focusing (IEF) with sodium dodecyl sulfate polyacrylamide gel electrophoresis (SDS-PAGE), resolves proteins by two independent physic principles: the isoelectric point (pI) in the first dimension and molecular mass (Mr) in the second dimension. Therefore, any modifications that may change protonable groups of a protein, such as those due to phosphorylation, acetylation, and alkylation, would greatly affect the migration rate on the first dimension, likewise modifications of molecular mass as glycosylation, ubiquitination, farnesylation, for example, might affect the migration rate on second dimension. A combination of these ptms might result in the vectorial addition of the two perpendicular shifts.

The result is a 2D gel map of resolved protein isoforms that may be visualized with either visible colorimetric method (Coomassie blue, silver stain) [14, 15] or fluorescent die [16]. In theory, 2DE is capable of detecting and quantifying protein amounts of less than 1 ng per spot.

After 2DE separation, gel images may be acquired using densitometers, and quantified by means of specific software for image analysis. After that, spots of interest are excised, destained, and analyzed by LC-MS/MS [8, 13, 17].

Here we give an overview of the technique and provide protocol useful for the analysis of the proteome of specific cellular line.

2. Protocol text

2.1 Methods

2.1.1 Protein isolation

Cell lines were washed twice with PBS (1X), PBS was carefully removed, and cells were scraped after addition of 500 μl of lysis buffer prepared in accordance with **Table 1**. Cell suspension was collected, incubated at 0°C for 30 min, and sonicated using Ultrasonic Baths (VWR) at 4°C for 10 s. Cell lysate was centrifuged at 15000×*g* for 20 min. The supernatant was carefully removed and stored at −80°C for further analysis [9].

2.1.2 Protein concentration assay

The main step of proteomic analysis is the measurement of protein extract concentration. An accurate protein quantitation is essential to avoid bias in the determination of differentially expressed proteins.

Protein content from control and treated samples was measured using the Bradford method (Bio-Rad) [18].

Standard curve preparation (for sample with 1–18 μg ml^{-1} protein)

We prepare seven dilutions of a HSA protein standard in a range from 1 to 18 μg protein.

A total of 10 μl of each standard solution was diluted in 790 μl of Milli-Q water. A total of 200 μl of Bradford reagent was added to each tube; the solution was mixed thoroughly and incubated at RT for 15 min in the dark.

Buffer	Components	Notes
Lysis buffer (LB)	15 mM Tris HCl pH 7.5, 120 mM sodium chloride, 30 mM potassium chloride, dithiothreitol 0.1%, and Triton X-100, 0.5%, supplemented with protease and phosphatase inhibitor cocktail (Halt Protease Inhibitor Cocktail/Halt Phosphatase Inhibitor Cocktail, Thermo Fisher Scientific Inc.).	Add protease inhibitors just prior to use. The 2X lysis buffer can be stored at 4°C up to 1 month.
Rehydration buffer (RB)	8 M Urea, 4% CHAPS, 0.8% (v/v) carrier ampholytes (pH 3–10 NL), traces of bromophenol blue, and 70 mM DTE (add as powder).	Add DTE and carrier ampholytes just prior to use. RB buffer can be stored at −20°C up to 6 months.
Equilibration buffer A	6 M Urea, 2% (w/v/) SDS, 50 mM Tris–HCl, pH 8.8, 30% (v/v), and 1% (w/v) dithiothreitol (DTT).	Dissolve by shaking, but not by heating the solution. All chemical used must be of ultrapure quality.
Equilibration buffer A	6 M Urea, 2% (w/v/) SDS, 50 mM Tris–HCl, pH 8.8, 30% (v/v), 2.5% (w/v) iodoacetamide (IAA).	Dissolve by shaking, but not by heating the solution. All chemical used must be of ultrapure quality.
Acrylamide stock	30% acrylamide–bis solution (29:1).	
SDS stock	20% SDS in MQ-water.	
Tris stock	1.5 M Tris–HCl, pH 8.8.	
APS stock	10% ammonium persulfate.	Prepare fresh by dissolving 0.1 g APS in 1 mL MQ-water.
SDS running buffer (lower chamber)	25 mM Tris base, 192 mM glycine, and 0.1% SDS.	
SDS running buffer (upper chamber)	25 mM Tris base, 192 mM glycine, and 0.2% SDS.	
Agarose sealing solution	1% Agarose in 1X SDS running buffer with trace of bromophenol blue.	
Gel solution	10% acrylamide–bis solution (29:1), 37.5 M Tris–HCl, pH 8.8, $Na_2S_2O_3$ 0.025% , APS 0.1%, and TEMED 0.04%; adjust volume using MQ-water.	Add TEMED and APS just before use.
Fixing solution (silver staining)	40% (w/v) Ethanol and 10% (w/v) acetic acid.	Prepare the solution using Milli-Q water.
Sensiting solution (silver staining)	0.02% Sodium thiosulfate pentahydrate.	Prepare the solution using Milli-Q water.
Silver nitrate solution (silver staining)	0.2% Silver nitrate and 0.02% formaldehyde (37%).	Prepare the solution using Milli-Q water.
Developing solution (silver staining)	3% Sodium carbonate, 0.05% formaldehyde (37%), and 0.005% sodium thiosulfate pentahydrate.	Prepare the solution using Milli-Q water. Solution must be prepared just before use.
Stop solution (silver staining)	2% Acetic acid.	Prepare the solution using Milli-Q water.

Buffer	Components	Notes
Destaining solution (Spot excision)	30 mM Potassium ferricyanide and 100 mM sodium thiosulfate.	Prepare the solution using HPLC grade water.
Gel washing solution (spot excision)	50% acetonitrile, 50 mM ammonium bicarbonate.	Prepare the solution using HPLC grade water.
Digestion buffer (spot excision)	Resuspend lyophilized trypsin (20 μg/vial) in 100 μL of the 1 mM HCl, yielding a 0.2 μg/μL stock solution. Aliquot into 10 × 10 μl and place at −20°C. To obtain the working solution, take 10 μl of this trypsin solution and add 500 μl of 50 mM NH_4HCO_3.	Prepare the solution using HPLC grade water. Avoid repeated freeze-thaw of trypsin stock solutions more than once.
Extraction solution	50% ACN–0.2% trifluoroacetic acid (TFA).	

Components and notes for buffers described and used in this chapter are summarized.

Table 1.
Buffer components.

The absorbance was measured at 595 nm using a BioPhotometer (Eppendorf). The assay was done in triplicate using a specific clear cuvette (UV-Vis transparent between 220 and 1600 nm). The BioPhotometer was calibrated on the basis of standard quantitative data.

About 2 μl of each protein extract was suspended in 798 μl of Milli-Q water, 200 μl of Bradford reagent was added to the mixture, vortexed, and incubated at RT for 15 min in the dark. The assay was performed in triplicate. Sample absorbance was measured at 595 nm using a calibrated BioPhotometer (Eppendorf).

2.1.3 First dimension: isoelectric focusing (IEF)

During isoelectric focusing, proteins are resolved in function of their isoelectric point (pI). The pI of a protein is the value of pH at which the electrophoretic mobility is nil. The majority of proteins have a pI within pH 4–7, although some proteins show an extreme value of pI as histone proteins. IEF is carried out using a precast strip of polyacrylamide with the most appropriate pH gradient range. The use of immobilized pH gradient (IPG) strips improves reproducibility, resolution, and protein load capability.

The pH gradient is created by dissolving in the IPG a mixture of aliphatic polyamino-polycarboxylic acids (carrier ampholines) that under the influence of an electric field create a stable pH gradient.

Proteins are resolved, in addition to their amino and carboxyl group, in function of a protonable group in their side chains. When the pH < pI, a protein has a net positive charge and at pH > pI, it becomes negatively charged. Therefore, under the influence of an electric field, proteins with acidic pI migrate toward the cathode and those with basic pI move toward the anode, until it reaches the strip region in which pH = pI. The method allows the separation of all protein isoforms [19].

In this experiment, we loaded 130 μg of each sample diluted into RB buffer (**Table 1**) during the rehydration step. When IEF is carried out for comparative experiments, IPG strips must have the same length and identical pH range; samples must be handled in parallel.

The rehydration volume is a function of strip length [19], for a strip 24 cm long, the appropriate volume is 480 μl. The rehydration tray was cleaned, dried, and leveled to ensure uniform reswelling.

A total of 480 μl of RB buffer, containing the sample, was distributed along the reswelling slot. The strips were deprived of protective cover and carefully flat on the fluid with the gel side down. The strips were checked for the presence of bubbles below and covered with mineral oil to prevent sample evaporation and urea crystallization. The tray was closed with the lid and rehydration was carried out for 16 h at RT.

IEF conditions may vary accordingly with sample amount, sample composition, IPG strip length, and pH range.

After the rehydration step, the IPG strips were removed from the slot, rinsed with Milli-Q water, and transferred on the Ettan IPGphor ceramic tray. During IEF, the gel strip is sided up. The strips were overlaid with mineral oil and the movable electrodes were assembled properly at the cathodic and the anodic end of the strip. A bit piece of filter paper was positioned between the strip and the electrode to eliminate ion excess.

IEF was carried out on a GE Healthcare IPGphor unit at 20°C, applying 50 uA/Strip, until a total of 70,000 Vh was reached. The following IPGphor program was used: (1) 300 V 3 h step, (2) 1000 V 6 h gradient, (3) 8000 V 3 h gradient, and (4) 8000 V 3 h step.

2.1.4 Second dimension

During SDS PAGE, IPG-focused proteins were separated according to their molecular weights.

For gel casting, a single gel for each IPG strip was prepared. Gel casting was done simultaneously. Gel caster was assembled with glass plates, separator sheets, and blank cassette inserts. Each component was previously cleaned and dried. Monomer solution was prepared according to the protocol (**Table 1**: gel solution) and carefully poured in the funnel casting.

After 2 min, the surface of each gel was layered with 2 ml of isopropanol to minimize exposure to oxygen. After 10 min, isopropanol was replaced with Milli-Q water and the gel was allowed to polymerize for at least 5 h. The gels were prepared concomitantly with the start of IEF and stored at 4°C until use.

Equilibration step: the equilibration step is essential to saturate focused proteins with SDS and allow their separation by SDS PAGE.

After IEF, IPG strips were equilibrated using 15 ml of equilibration buffer A (**Table 1**) followed by treatment with 15 ml of equilibration buffer B (**Table 1**) [19, 20]. The equilibrated IPG strip was rinsed with running buffer 1X and positioned with acidic end to the left and the plastic back facing one the higher glass plates, avoiding the formation of bubbles between the strip and the top of the gel.

A small piece of filter paper soaked with 10 μL of protein MW marker was positioned at the right side using forceps.

The strips were sealed using agarose solution. The second dimension was done on 10% SDS-polyacrylamide gels.

The electrophoretic unit was filled with running buffer 1X and connected with the MultiTemp III Thermostatic Circulator. When the temperature of the buffer reached the seat value of 25°C, the gels were positioned in the chamber; the empty slots were filled with blank cassettes. The upper chamber was positioned in the electrophoretic unit and filled with running buffer 1X, SDS 0.02%. The lid was closed and attached to the power supply. Run was done at constant voltage (2 W/gel, 25°C) until the bromophenol blue dye front reached the end of the gels [20, 21].

After electrophoresis, gels were carefully removed from their gel cassettes and placed on a glass tray for staining.

2.1.5 Staining with MS-compatible silver staining

Gels were stained with the MS-compatible silver staining procedure [14, 21]. The methods used for staining are summarized in **Table 2.**

2.1.6 Gel image analysis

The main aim of proteomics consists in the identification of differentially expressed proteins. To proceed with the 2DE analysis, the first step is the digitalization of gel images. It has been done by Image Scanner II (GE healthcare).

The gels were scanned at 600 dpi and saved as tiff.

Subsequent analysis was performed using the Image Master 2D-Platinum software, version 6.0 (GE Healthcare) according to the software manual. Briefly, raw images were imported into the software and processed for protein detection. The spot auto-detect function was used for all group comparisons applying the following parameters: smooth 2, min area 5, saliency 1.00000. Digital images were checked to ensure that all the proteins present in the gel were correctly detected. Differences in protein expression were assessed using the relative volume (%Vol) option of the software. Using this option, the data becomes independent variables caused by differences in loading or staining [22].

Analysis was done using three independent replicates. All data were presented as mean ± SEM (N), where SEM is the standard error of the mean and N represents the number of replicates. Protein levels in every data set were compared to control group using unpaired t-test. We considered statistically significant a two-sided p-value <0.05. Excel spreadsheet (Microsoft) was used for data plotting [23].

2.1.7 Spot excision and digestion

A main issue during spot excision and digestion procedure is avoiding contamination with keratins that can affect subsequent mass spectrometry analysis. Proper lab-equipment and dedicated reagents mast be used for protein spot

Step	Solution	Volume	Time
Fixing	Fixing solution	500 ml	2 h
Washing	Wash with 30% ethanol	500 ml	30 min
Sensiting	Sensiting solution	500 ml	1 min
Washing	Milli-Q water	500 ml	2 × 2 min
Staining	Silver nitrate solution	500 ml	30 min
Rinsing	Milli-Q water	500 ml	2 × 2 min
Developing	Developing solution	500 ml	2 min
Developing	Developing solution	500 ml	Develop for the time necessary to clearly visualize the spot.
Stop solution	Stop solution	500 ml	20 min
Washing	Milli-Q water	500 ml	3 × 10 min

The table summarizes the steps used for silver staining.

Table 2.
Silver staining steps.

digestion. All instruments, as well as the benchtop, were meticulously cleaned with 70% ethanol.

Protein spots, with a differential expression profile, were manually excised from silver stained gel using OneTouch Plus Spotpicker (Gel company).

Gel pieces were washed with Milli-Q water and destained incubating with a destaining solution (**Table 1**) for 15 min.

The destaning solution was removed and gel spot washed three times with 200 μl of 200 mM NH_4HCO_3 pH 7.8 for 15 min; the incubation was done in a tube shaker at 37°C. Spots were incubated with ACN for 10 min at 37°C. A total of 70 μl of trypsin working solution (**Table 1**) was added to gel pieces prior to incubation at 37°C overnight. Trypsin solution was carefully recovered and placed in a fresh tube (t).

Additional peptides were recovered by incubating the gel piece with extraction solution at 37°C for 10 min. The solution was recuperated and pooled in the fresh tube (t). Peptide solution was concentrated using a speedy vac [24, 25].

The resulting tryptic peptides were purified by Pierce® C18 Spin Columns (Thermo Fisher Scientific Inc.) according to the manufacturer's procedure, eluted with 40 μL of 70% acetonitrile, and dehydrated in a vacuum evaporator [26].

2.1.7.1 Nanoscale LC-MS/MS analysis

Tryptic peptides were analyzed using Nanoscale LC-MS/MS. The analysis was performed as service in our mass spectrometry facility.

Chromatography analysis was done on Easy LC 1000 nanoscale liquid chromatography (nanoLC) system (Thermo Fisher Scientific, Odense, Denmark) as previously reported by Scumaci et al. [23]. Purified peptides were re-suspended with 0.1% formic acid and injected at 500 nL/min on the analytical column. A binary gradient was used for the elution of peptides. Mass analysis was done by means of a quadrupole-orbitrap mass spectrometer Q-Exactive (Thermo Fisher Scientific, Bremen, Germany) working in positive ion mode, using nanoelectrospray (nESI) potential at 1800 V. Acquisition of data was achieved with a top-5 method using resolution (FWHM), AGC target, and maximum injection time (ms) for full MS and MS/MS of, respectively, 70,000/17,500, 1e6/5e5, and 50/400. Precursor ion isolation was done using a mass window of 2.0 m/z, while normalized collision energy was 30. Ion threshold for generating MS/MS events was 2e4. Dynamic exclusion was set at 15 s. Data were processed using Proteome Discoverer 1.3 (Thermo Fisher Scientific, Bremen, Germany), search engine was Sequest, and the HUMAN-refprot-isoforms.fasta was used as a sequence database. Searching parameters were: MS tolerance 15 ppm; MS/MS tolerance 0.02 Da; fixed modifications: cysteine carbamidomethylation; variable modification: methionine oxidation, serine, threonine, and tyrosine phosphorylation; enzyme: trypsin; max. missed cleavages 2; taxonomy Human. We consider only protein hits generated with two successful peptide identifications valid and acceptable [27].

2.1.8 Western blot validation

Western blot analysis of prohibitin and gelsolin was done to verify 2D gel electrophoresis data. Analysis was performed on cell extracts of MCF7wt and BRCA1 interfered MCF7.

About 50 μg of each protein extract were resolved on a 4–15% SDS PAGE precast gel (Bio-Rad), and blotted to a nitrocellulose membrane followed by immunoblotting. The working concentration of rabbit monoclonal antibody against BRCA1 (clone D-20, Santa Cruz was 1 μg/ml) [23].

Anti-PHB1 (Prohibitin 1) from cell signaling technology was used at 1:1000 dilution and Anti-Gelsolin (Santa Cruz) was used at 1:1000 dilution. The signal was detected using the specific horseradish peroxidase-conjugate secondary antibodies; blots were developed using the SuperSignal West Femto ECL substrate (Pierce). Densitometric software (Alliance 2.7 1D fully automated software) determined the percent distribution of blotted proteins after image acquisition by Alliance 2.7 (UVITEC, Eppendorf, Milan, Italy) [28].

Data were analyzed and plotted using Excel spreadsheet (Microsoft), and expressed as mean ± SEM (N), where SEM represents the standard error of the mean and N indicates the number of experimental repeats. Unpaired t-test was used to compare protein levels in each data set. A two-sided p-value <0.05 was considered statistically significant.

3. Results

2D gel electrophoresis was applied to analyze the whole proteome of breast cancer cell expressing or not BRCA1. BRCA1 is a tumor suppressor gene, often mutated in hereditary breast and ovarian cancers. The gene encodes for a large protein involved in several cellular pathways that include DNA damage-induced cell cycle checkpoint activation, maintenance of genomic stability, DNA damage repair, as well as chromatin remodeling, protein ubiquitination, transcriptional regulation, and apoptosis [29]. Recently, it has been also reported that BRCA1 is able to induce reprogramming of metabolism toward aerobic glycolysis in breast cancer cells [30]. Here we induce the transient inactivation of BRCA1 in MCF-7 cell line for 48 h by transfection with Oligo siRNA/Brca1 [20, 31]. It has been done by Lipofectamine 2000 reagent (Invitrogen, Paisley, UK), according to the manufacturer's instructions. Oligo siRNA/Brca1 duplex was purchased from Sigma Aldrich and used at a final concentration of 100 nM. The silencing of BRCA1 was checked by western blot analysis (**Figure 1**). Transient knock down of BRCA1 was done at 48 h according to our previous results [20]. The analysis workflow is summarized in **Figure 2**. The analysis was performed using three biological replicates.

Using 2D gel electrophoresis, we obtained a gel map of about 500 protein spot for each replicate (**Figure 3**). Image analysis allowed to identify 25 differentially

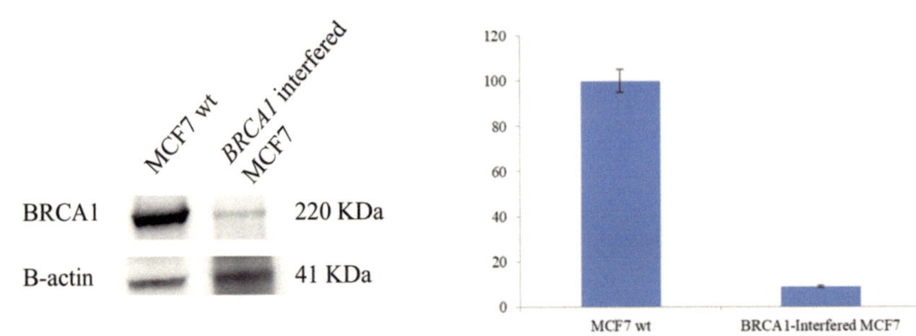

Figure 1.
Gene interference. Western blot analysis confirmed BRCA1 switch off. Equal amounts of MCF7 protein extracts were separated on a 4–15% SDS PAGE precast gel (Bio-Rad) and transferred to a nitrocellulose membrane followed by immunoblotting. In each panel, the left panel is representative data of western blot analysis; the right panel shows densitometric analysis. Analysis was performed using three independent experiments. Data are mean ± SEM (N = 3). $p < 0.05$. In each panel, β-actin blot shows equal amounts of protein loading.

expressed proteins. Of these, 15 protein spots were down-expressed, while 10 were up-regulated.

Differentially expressed proteins were manually excised, destained, trypsin digested, and identified using LC-MS/MS analysis. A list of identified protein is provided in **Table 3**.

2DE data were confirmed using western blot analysis for two selected candidates (**Figure 4**).

Differentially expressed protein was analyzed using Ingenuity software PA tool.

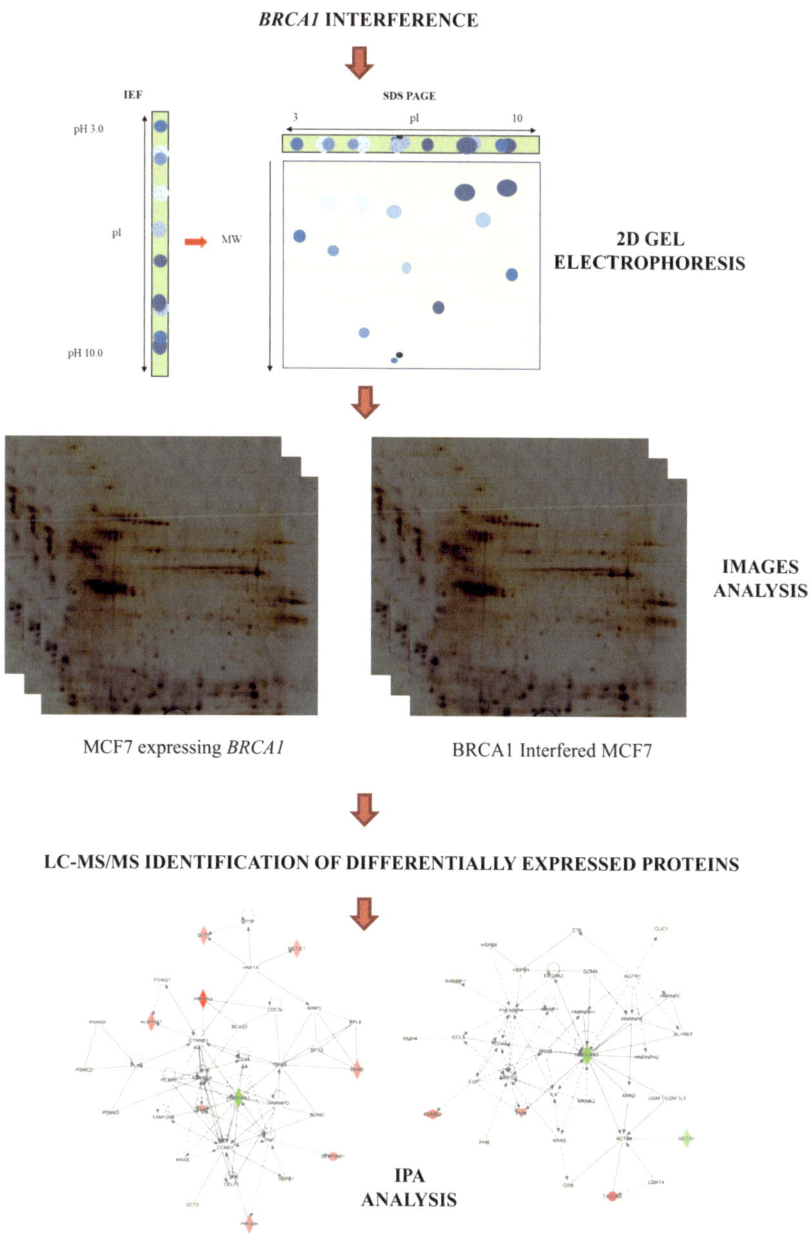

Figure 2.
Proteomic approach work flow.

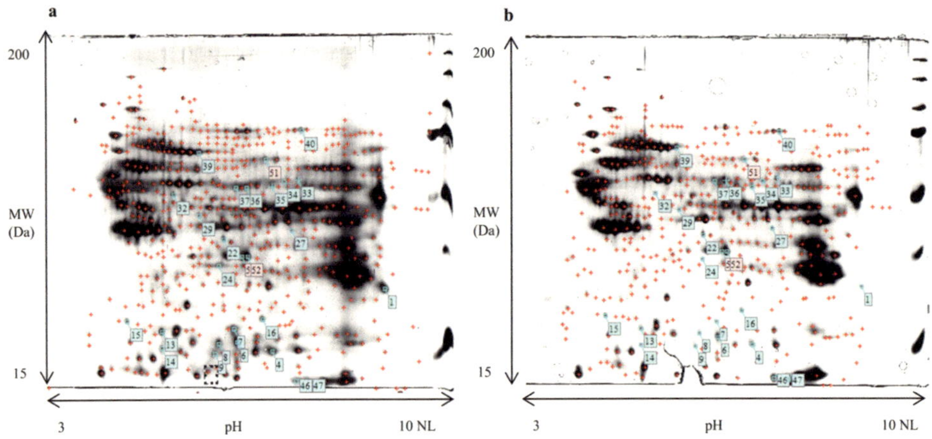

Figure 3.
2D gel electrophoresis maps. (a) Representative 2D gel electrophoresis maps of MCF7 wt protein extracts and (b) representative 2D gel electrophoresis maps of BRCA1 interfered MCF7 protein extracts. Isoelectrofocusing was carried out on 3–10 NL IPGstrip, 24 cm length. The second dimension was performed on 10% SDS-PAGE. Gel images were analyzed using Decider software. Numbered spots indicate proteins that have statistically significant differential expression between samples according to Image master 2D Platinum 7.0 software (GE Healthcare).

Spot ID	ID protein	Description	Score	Coverage	Unique peptides	Molecular weight	pI	BRCA1 interfered MCF7/ MCF7 wt
1	P09651-3	Isoform 2 of Heterogeneous nuclear ribonucleoprotein A1 OS=*Homo sapiens* GN=HNRNPA1 – [ROA1_HUMAN]	76.57	43.07	9.00	29.37	9.14	−24.63
4	Q15102	Platelet-activating factor acetylhydrolase IB subunit gamma OS=*Homo sapiens* GN=PAFAH1B3 PE=1 SV=1 – [PA1B3_HUMAN]	5.59	11.26	3.00	25.72	6.84	−2.60
6	P25788-2	Isoform 2 of Proteasome subunit alpha type-3 OS=*Homo sapiens* GN=PSMA3 – [PSA3_HUMAN]	18.38	23.79	7.00	27.63	5.33	−1.93
7	P78417	Glutathione S-transferase omega-1 OS=*Homo sapiens* GN=GSTO1 PE=1 SV=2 – [GSTO1_HUMAN]	14.83	21.99	6.00	27.55	6.60	−13.10

Spot ID	ID protein	Description	Score	Coverage	Unique peptides	Molecular weight	pI	BRCA1 interfered MCF7/ MCF7 wt
8	P43487	Ran-specific GTPase-activating protein OS=*Homo sapiens* GN=RANBP1 PE=1 SV=1 – [RANG_HUMAN]	0.00	8.96	2.00	23.30	5.29	−3.84
9	P04792	Heat shock protein bcta-1 OS=*Homo sapiens* GN=HSPB1 PE=1 SV=2 – [HSPB1_HUMAN]	22.60	30.73	6.00	22.77	6.40	−4.42
13	P35232	Prohibitin OS=*Homo sapiens* GN=PHB PE=1 SV=1 – [PHB_HUMAN]	46.62	45.22	11.00	29.79	5.76	−1.94
14	P43487	Ran-specific GTPase-activating protein OS=*Homo sapiens* GN=RANBP1 PE=1 SV=1 – [RANG_HUMAN]	4.76	8.96	2.00	23.30	5.29	−4.99
15	O00299	Chloride intracellular channel protein 1 OS=*Homo sapiens* GN=CLIC1 PE=1 SV=4 – [CLIC1_HUMAN]	55.50	57.68	9.00	26.91	5.17	−1.15
16	P00491	Purine nucleoside phosphorylase OS=*Homo sapiens* GN=PNP PE=1 SV=2 – [PNPH_HUMAN]	10.39	16.96	4.00	32.10	6.95	−2.77
22	P30740	Leukocyte elastase inhibitor OS=*Homo sapiens* GN=SERPINB1 PE=1 SV=1 – [ILEU_HUMAN]	25.33	27.70	10.00	42.71	6.28	2.21
24	Q9NP79	Vacuolar protein sorting-associated protein VTA1 homolog OS=*Homo sapiens* GN=VTA1 PE=1 SV=1 – [VTA1_HUMAN]	4.65	9.45	2.00	33.86	6.29	−4.32

Spot ID	ID protein	Description	Score	Coverage	Unique peptides	Molecular weight	pI	BRCA1 interfered MCF7/ MCF7 wt
27	B7Z7E9	Aspartate aminotransferase OS=*Homo sapiens* GN=GOT1 PE=2 SV=1 – [B7Z7E9_ HUMAN]	3.86	8.67	3.00	44.12	8.07	2.02
29	P35998	26S protease regulatory subunit 7 OS=*Homo sapiens* GN=PSMC2 PE=1 SV=3 –[PRS7_HUMAN]	93.17	60.28	26.00	48.60	5.95	−1.45
32	F8VPV9	ATP synthase subunit beta OS=*Homo sapiens* GN=ATP5B PE=2 SV=1 – [F8VPV9_ HUMAN]	20.60	22.39	8.00	55.27	5.40	−3.96
33	P14618	Pyruvate kinase isozymes M1/M2 OS=*Homo sapiens* GN=PKM PE=1 SV=4 – [KPYM_ HUMAN]	29.08	32.02	13.00	57.90	7.84	2.36
34	P00352	Retinal dehydrogenase 1 OS=*Homo sapiens* GN=ALDH1A1 PE=1 SV=2 – [AL1A1_HUMAN]	21.35	24.35	10.00	54.83	6.73	2.31
35	P14868	Aspartate--tRNA ligase, cytoplasmic OS=*Homo sapiens* GN=DARS PE=1 SV=2 – [SYDC_ HUMAN]	7.94	8.18	4.00	57.10	6.55	2.15
36	Q9UMS4	Pre-mRNA-processing factor 19 OS=*Homo sapiens* GN=PRPF19 PE=1 SV=1 – [PRP19_ HUMAN]	7.76	8.93	4.00	55.15	6.61	3.80
37	O43175	D-3-phosphoglycerate dehydrogenase OS=*Homo sapiens* GN=PHGDH PE=1 SV=4 – [SERA_HUMAN]	40.04	29.08	12.00	56.61	6.71	2.16
39	B4DUR8	T-complex protein 1 subunit gamma OS=*Homo sapiens* GN=CCT3 PE=2 SV=1 – [B4DUR8_ HUMAN]	11.04	15.20	6.00	55.64	5.64	−1.85

Spot ID	ID protein	Description	Score	Coverage	Unique peptides	Molecular weight	pI	BRCA1 interfered MCF7/ MCF7 wt
40	Q96RQ3	Methylcrotonyl-CoA carboxylase subunit alpha, mitochondrial OS=*Homo sapiens* GN=MCCC1 PE=1 SV=3 – [MCCA_HUMAN]	36.71	24.69	13.00	80.42	7.78	1.84
46	P49721	Proteasome subunit beta type-2 OS=*Homo sapiens* GN=PSMB2 PE=1 SV=1 – [PSB2_HUMAN]	14.09	27.86	5.00	22.82	7.02	2.90
47	P37802	Transgelin-2 OS=*Homo sapiens* GN=TAGLN2 PE=1 SV=3 – [TAGL2_HUMAN]	12.17	34.17	6.00	22.38	8.25	3.23
51	P13020-2	Isoform 2 of Gelsolin OS=*Mus musculus* GN=Gsn – [GELS_MOUSE]	10.66	4.10	3.00	80.71	5.72	−2.03
52	P60709	Actin, cytoplasmic 1 OS=*Homo sapiens* GN=ACTB PE=1 SV=1 – [ACTB_HUMAN]	13.39	17.07	2.00	41.71	5.48	−1.15

In the table are reported mass spectrometry identifications for protein spots that resulted differentially expressed following 2DE analysis. For each spot are reported: spot id, gene name, identification score, number of unique peptides, isolectric point (pI), and molecular weight (MW). In the last columns are reported quantitative data. Data are reported as fold change between BRCA1-interfered breast cancer cells (MCF7) versus breast cancer cell (MCF7) wt. (p value < 0.05).

Table 3.
Densitometric data of 2D gel electrophoresis and protein identification by LC-MS/MS analysis.

3.1 Pathway and network analysis

Differentially expressed proteins were connected using ingenuity pathway analysis (Ingenuity Systems, www.ingenuity.com). Data sets with protein identification and quantitative data were loaded into the application that creates hypothetical networks of protein interaction based on ingenuity pathway knowledge base. IPA software generates a list of networks in function on their connectivity, allowing to associate specific biological functions to genes here included [23, 27, 29, 32].

Proteins were mapped onto three networks (**Figure 5**). The most representative, with a score of 28 and 13 focus molecules, had function associated with cancer, cellular movement, and connective tissue disorders. The second network, including 12 focus molecules with a score of 25, exhibited functions connected with free radical scavenging, protein synthesis, DNA replication, recombination, and repair. Finally, the last network grouped proteins involved in cancer, cell death and survival, and cell morphology.

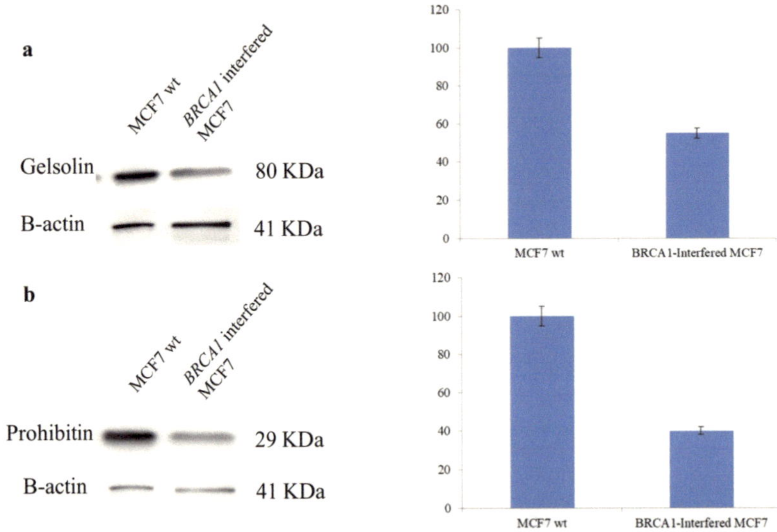

Figure 4.
Western blot analysis confirming 2DE data for gelsolin (a) and prohibitin (b). Analysis was performed on protein extracts from MCF7 expressing or not BRCA1. Equal amounts of MCF7 protein extracts were separated on a 4–15% SDS PAGE precast gel (Bio-Rad) and electrotransferred to a nitrocellulose membrane with a Trans-blot turbo system (Biorad). Images were acquired using the Alliance 2.7 system (UVITEC, Eppendorf, Milan, Italy) and analyzed by Excel spreadsheet. In each panel, the left panel is representative of western blot analysis; the right panel shows densitometric analysis. Analysis was performed using three independent experiments. Data are mean ± SEM (N = 3). $p < 0.05$. In each panel, β-actin blot shows equal amounts of protein loading.

a

ID	Molecules in Network	Score	Focus Molecules	Top Diseases and Functions
1	AIMP1,ALDH1A1,BCAS2,CCND1,CCT3,CD44,CDC5L,CDKN1A,CELF1,CTNNB1,DARS,FAM129B,FOXQ1,GOT1,GRB2,HNF1A,HNRNPA1,HNRNPD,HSPB1,KRAS,MCCC1,MTTP,NONO,PCBP1,PHGDH,PKM,PLK1,PRPF19,PSMA3,PSMA5,PSMC2,RPL8,RPS3,SERPINB1,TNF	28	13	Cancer, Cellular Movement, Connective Tissue Disorders
2	ACTB,AGTR1,ALYREF,CCL5,CD44,CDK14,CLIC1,EGF,EIF2AK2,ERK1/2,GSN,GSTO1,GZMA,HNRNPA1,HNRNPC,HNRNPH1,HNRNPH2,HNRNPK,HRAS,HSPB1,HSPB8,ILK,KRAS,MKNK1,MKNK2,P38 MAPK,PHB,PKM,PNP,PSMB2,RANBP1,TAGLN2,TTN,U2AF1/U2AF1L5,XRN2	25	12	Free Radical Scavenging, Protein Synthesis, DNA Replication, Recombination, and Repair
3	SATB1,SUZ12,VTA1	2	1	Cancer, Cell Death and Survival, Cell Morphology

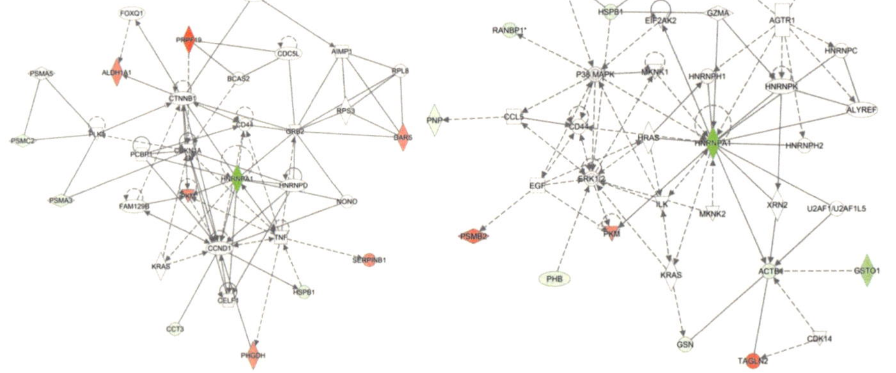

Figure 5.
IPA analysis. (a) the list of networks generated from IPA analysis and significantly modulated (log p-value) in BRCA1-interferd MCF7 versus MCF7 wt.; the top 2 signaling networks; (b) cancer, cellular movement, and connective tissue disorders; (c) free radical scavenging, protein synthesis, DNA replication, recombination, and repair.

4. Discussion

The aim of this work is to describe how 2D gel electrophoresis might be effectively applied to provide a comprehensive, quantitative description of protein expression and its change under the influence of a specific biological perturbation. Here we provide an overview of the 2DE methodologies with special focus on the technical procedure.

To provide an example of the technique application, we describe the proteome profiling of breast cancer cells following BRCA1 interference.

Using 2D gel electrophoresis, we found that BRCA1 silencing affected the expression of 25 proteins belonging to different cellular compartments and implicated in a variety of cellular pathways. We focused our attention on the downregulation of prohibitin and gelsolin as well as the upregulation of pyruvate kinase.

Upon BRCA1 interference, we found a reduced level of prohibitin, coherently with literature data that highlighted the ability of prohibitin in inhibiting breast cancer cell proliferation and promoting apoptosis [33]. In addition, we found a reduced level of gelsolin, fully in agreement with what we previously published [20]. Otherwise, the expression of pyruvate kinase, together with D-3-phosphoglycerate dehydrogenase, was found up-regulated, according with a metabolic shift toward glucose metabolism [34].

Although interesting, these findings deserve validations and further investigations that largely exceed the scope of this chapter. However, the approach provided the opportunity to highlight the capability of 2D gel electrophoresis in proteome profiling breast cancer cells and in detecting the specific phenotypic changes induced by the interference of the key gene BRCA1.

The fundamental step of our proteomic analysis was the preparation of the samples. Protein extracts for 2DE were done directly on a plate with a minimum amount of lysis buffer, to avoid contaminations with 2DE interfering substances, but enough to ensure the complete extraction, solubilization, and disaggregation of all cellular proteins avoiding proteolysis and poor recovery. This was a critical step because experimentally induced protein changes may be challenging to distinguish from biological differences under investigations. Sample protein content was carefully quantized and loaded using the "in-gel rehydration" procedure. The second dimension was performed at constant voltage and temperature. Silver staining was done in glass tray with accuracy, avoiding the formation of gray precipitates as well as the formation of the dark background caused by an extra developing.

The reproducibility of the spot pattern in gel images of different samples reflects the accurateness of the procedures. Moreover, the presence of spot well distributed on the gel with finely defined borders as well as the lack of streak and background staining proved the quality of 2D separation.

5. Conclusions

In this chapter, we point out the key aspects of 2D gel electrophoresis describing in practice the use of the technique for the proteomic profiling of breast cancer cells expressing or not BRCA1.

We offer an overview of methodological peculiarities, undelaying that, in spite of some technical limitations, 2D gel electrophoresis is still one of the major current analytical technologies useful to investigate specific change in the profiles of proteome in different cells, tissues, or organism [35].

The proteome of higher eukaryotes holds a huge complexity due to dynamic changes in protein expression, modification, and interactions. It's obvious that a single proteomic approach is not enough to address complex biological questions, but we are confident that the methods, here deepened, represents a gold standard for the analysis of intact proteins and all relative post-translational modifications responsible for a peculiar phenotype or function [36–40].

2DE is unrivaled by other proteomic approach because it consents the simultaneous detection and quantification of thousands of proteins isoforms, not predictable through other techniques including genome analysis [41, 42].

We are confident that protocols and methodologies we described are going to be useful for all scientists interested in 2DE separation to address their own biological questions.

Acknowledgements

The authors would like to acknowledge the Department of Experimental and Clinical Medicine (UMG) for support to Domenica Scumaci, and MIUR grant PON03PE_00009_2(iCARE) for support to Giovanni Cuda.

Conflict of interest

The authors declare that they have no competing financial interests.

Author details

Domenica Scumaci* and Giovanni Cuda
Laboratory of Proteomics, Research Center on Advanced Biochemistry and Molecular Biology, Department of Experimental and Clinical Medicine, Magna Græcia University of Catanzaro, Italy

*Address all correspondence to: scumaci@unicz.it

References

[1] Wasinger VC, Cordwell SJ, Cepa-Poljak A, Yan JX, Gooley AA, Wilkins MR, et al. Progress with gene-product mapping of the Mollicutes: Mycoplasma genitalium. Electrophoresis. 1995;**16**:1090

[2] Wilkins MR, Sanchez JC, Gooley AA, Appel RD, Humphery-Smith I, Hochstrasser DF, et al. Progress with proteome projects: Why all proteins expressed by a genome should be identified and how to do it. Biotechnology & Genetic Engineering Reviews. 1996;**13**:19-50. Review

[3] Fields S. Proteomics in Genomeland. Science. 2001;**291**(5507):1221-1224. DOI: 10.1126/science.291.5507.1221

[4] O'Farrell PH. High resolution two-dimensional electrophoresis of proteins. The Journal of Biological Chemistry. 1975;**250**(10):4007-4021

[5] Wittmann-Liebold B, Graack HR, Pohl T. Two-dimensional gel electrophoresis as tool for proteomics studies in combination with protein identification by mass spectrometry. Proteomics. 2006;**6**(17):4688-4703

[6] Yoo DY, Cho SB, Jung HY, Kim W, Choi GM, Won MH, et al. Moon SM6Tat-protein disulfide-isomerase A3: A possible candidate for preventing ischemic damage in the spinal cord. Cell Death & Disease. 2017;**8**(10):e3075. DOI: 10.1038/cddis.2017.473

[7] Pietrovito L, Cano-Cortés V, Gamberi T, Magherini F, Bianchi L, Bini L, et al. Cellular response to empty and palladium-conjugated amino-polystyrene nanospheres uptake: A proteomic study. Proteomics. 2015;**15**(1):34-43. DOI: 10.1002/pmic.201300423

[8] Morelli M, Scumaci D, Di Cello A, Venturella R, Donato G, Faniello MC, et al. DJ-1 in endometrial cancer: A possible biomarker to improve differential diagnosis between subtypes. International Journal of Gynecological Cancer. 2014;**24**(4):649-658. DOI: 10.1097/IGC.0000000000000

[9] Amato R, Scumaci D, D'Antona L, Iuliano R, Menniti M, Di Sanzo M, et al. Sgk1 enhances RANBP1 transcript levels and decreases taxol sensitivity in RKO colon carcinoma cells. Oncogene. 2013;**32**(38):4572-4578. DOI: 10.1038/onc.2012.470

[10] Görg A, Obermaier C, Boguth G, Harder A, Scheibe B, Wildgruber R, et al. The current state of two-dimensional electrophoresis with immobilized pH gradients. Electrophoresis. 2000;**21**:1037-1053

[11] Barnouin K. Two-dimensional gel electrophoresis for analysis of protein complexes. In: Fu H, editor. Protein-Protein Interactions: Methods and Applications. Vol. 261. New Jersey: Humana Press Inc; 2004

[12] Rabilloud T, Chevallet M, Luche S, Lelong C. Two-dimensional gel electrophoresis in proteomics: Past, present and future. Journal of Proteomics. 2010;**73**(11):2064-2077. DOI: 10.1016/j.jprot.2010.05.016

[13] Rabilloud T, Lelong C. Two-dimensional gel electrophoresis in proteomics: A tutorial. Journal of Proteomics. 2011;**74**:1829-1841

[14] Shevchenko A, Wilm M, Vorm O, Mann M. Mass spectrometric sequencing of proteins silver-stained polyacrylamide gels. Analytical Chemistry. 1996;**68**(5):850-858

[15] Stensballe A, Jensen ON. Simplified sample preparation method for protein identification by matrix-assisted

laser desorption/ionization mass spectrometry: In-gel digestion on the probe surface. Proteomics. 2001;**1**(8):955-966

[16] Marouga R, David S, Hawkins E. The development of the DIGE system: 2D fluorescence difference gel analysis technology. Analytical and Bioanalytical Chemistry. 2005;**382**(3):669-678. Epub 2005 May 18. Review

[17] Wilkins MR, Pasquali C, Appel RD, Ou K, Golaz O, Sanchez JC, et al. From proteins to proteomes: Large scale protein identification by two-dimensional electrophoresis and amino acid analysis. Biotechnology (N. Y). 1996;**14**(1):61-65

[18] Bradford MM. A rapid and sensitive method for the quantitation of microgram quantities of protein utilizing the principle of protein-dye binding. Analytical Biochemistry. 1976;**72**:248-254. PMID: 942051

[19] Bjellqvist B, Ek K, Righetti PG, Gianazza E, Görg A, Westermeier R, et al. isoelectric focusing in immobilized pH gradients: Principle, methodology and some applications. Journal of Biochemical and Biophysical Methods. 1982;**6**(4):317

[20] Seillier-Moiseiwitsch F. The Proteomics Protocols Handbook. In: John M. Walker, editor. Humana Press. 2005:239-258. ISBN 978-1-59259-890-8. DOI: 10.1385/1592598900

[21] Weiss W, Gorg A. High-resolution two-dimensional electrophoresis. Methods in Molecular Biology. 2009;**564**:13-32. DOI: 10.1007/978-1-60761-157-8_2. PMID: 19544015

[22] Zivy M. Quantitative analysis of 2D gels. Methods in Molecular Biology. 2007;**355**:175-194

[23] Scumaci D, Tammè L, Fiumara CV, Pappaianni G, Concolino A, Leone E, et al. Plasma proteomic profiling in hereditary breast cancer reveals a BRCA1-specific signature: Diagnostic and functional implications. PLoS One. 2015;**10**(6):e0129762

[24] Hellman U, Wernsted C, Gonez J, Heldin CH. Improvement of an "In-Gel" digestion procedure for the micropreparation of internal protein fragments for amino acid sequencing. Analytical Biochemistry. 1995;**224**: 451-455. PMID: 7710111

[25] Gharahdaghi F, Weinberg CR, Meagher DA, Imai BS, Mische SM. Mass spectrometry identification of proteins from silver stained polyacrylamide gel: A method for the removal of silver ions to enhance sensitivity. Electrophoresis. 1999;**20**:601-605. PMID: 10217175

[26] Rappsilber J, Ishihama Y, Mann M. Stop and go extraction tips for matrix-assisted laser desorption/ionization, nanoelectrospray, and LC/MS sample pretreatment in proteomics. Analytical Chemistry. 2003;**75**:663

[27] Concolino A, Olivo E, Tammè L, Fiumara CV, De Angelis MT, Quaresima B, et al. Proteomics analysis to assess the role of Mitochondria in BRCA1-mediated breast tumorigenesis. Proteome. 2018;**6**(2):pii: E16. DOI: 10.3390/proteomes6020016

[28] Scumaci D, Oliva A, Concolino A, Curcio A, Fiumara CV, Tammè L, et al. Integration of "Omics" strategies for biomarkers discovery and for the elucidation of molecular mechanisms underlying Brugada syndrome. Proteomics Clinical Applications. Nov 2018;**12**(6):e1800065. DOI: 10.1002/prca.201800065. Epub 2018 Jul 20

[29] King MC. Insights into the functions of BRCA1 and BRCA2. Trends in Genetics. 2000;**16**:69-74

[30] Privat M, Radosevic-Robin N, Aubel C, Cayre A, Penault-Llorca F,

Marceau G, et al. BRCA1 induces major energetic metabolism reprogramming in breast cancer cells. PLoS One. 2014;**9**(7):e102438. DOI: 10.1371/journal.pone.0102438

[31] Bartz SR, Zhang Z, Burchard J, Imakura M, Martin M, Palmieri A, et al. Small interfering RNA screens reveal enhanced cisplatin cytotoxicity in tumor cells having both BRCA network and TP53 disruptions. Molecular and Cellular Biology. 2006;**26**(24):9377-9386

[32] Krämer A, Green J, Pollard J, Tugendreich S. Causal analysis approaches in ingenuity pathway analysis (IPA). Bioinformatics. 15 Feb 2014;**30**(4):523-530. DOI: 10.1093/bioinformatics/btt703. Epub 2013 Dec 13

[33] Liu P, Xu Y, Zhang W, Li Y, Tang L, Chen W, et al. Prohibitin promotes androgen receptor activation in ER-positive breast cancer. Cell Cycle. 2017;**16**(8):776-784. DOI: 10.1080/15384101.2017.1295193. Epub 2017 Mar 8

[34] Iqbal MA, Gupta V, Gopinath P, Mazurek S, Bamezai RN. Pyruvate kinase M2 and cancer: An updated assessment. FEBS Letters. 2014;**588**(16):2685-2692. DOI: 10.1016/j.febslet.2014.04.011. Epub 2014 Apr 18

[35] Gorg A, Weiss W, Dunn MJ. Current two-dimensional electrophoresis technology for proteomics. Proteomics. 2004;**4**:3665-3685

[36] Hartinger J, Stenius K, Hogemann D, Jahn R. 16-BAC/SDS-PAGE: A two-22 dimensional gel electrophoresis system suitable for the separation of integral membrane proteins. Analytical Biochemistry. 1996;**240**:126-133

[37] Burre J, Beckhaus T, Schagger H, Corvey C, Hofmann S, Karas M, et al. Analysis of the synaptic vesicle proteome using three gel-based protein separation techniques. Proteomics. 2006;**6**:6250-6262

[38] Yang WH, Kim JE, Nam HW, Ju JW, Kim HS, Kim YS, et al. Modification of p53 with O-linked N-acetylglucosamine regulates p53 activity and stability. Nature Cell Biology. 2006;**8**:1074-1083

[39] Seo J, Jeong J, Kim YM, Hwang N, Paek E, Lee KJ. Strategy for comprehensive identification of post- cellular proteins, including low abundant modifications: Glyceraldehyde-3-phosphate dehydrogenase. Journal of Proteome Research. 2008;**7**:587-602

[40] John JP, Pollak A, Lubec G. Complete sequencing and oxidative modification of manganese superoxide in medulloblastoma cells. Electrophoresis. 2009;**30**:3006-3016

[41] Maltese WA. Posttranslational modification of proteins by isoprenoids in mammalian-cells. The FASEB Journal. 1990;**4**:3319-3328

[42] Kim YI, Cho JY. Gel-based proteomics in disease research: Is it still valuable? Biochimica et Biophysica Acta - Proteins & Proteomics. 2019;**1867**(1):9-16. DOI: 10.1016/j.bbapap.2018.08.001. Epub 2018 Aug 15

Chapter 2

Precipitation of Detergent-Containing Samples for Top-Down and Bottom-Up Proteomics

Alan Doucette and Andrew Crowell

Abstract

Prior to proteome analysis by mass spectrometry (MS), protein mixtures must first be subject to various sample preparation steps. The goal is to isolate proteins in high yield, and with high purity. Liquid chromatography (LC) separation is also integral to comprehensive proteome characterization, and so a key component of sample preparation is simply to solubilize the proteome in LC-MS compatible solvents. Hydrophobic proteins (membrane proteins) represent a greater challenge to maintain protein solubility during sample preparation. Sodium dodecyl sulfate (SDS) is a favored detergent to solubilize proteins, and also is used to impart mass-based fractionation (i.e., SDS PAGE, GELFrEE). However, SDS is incompatible with downstream LC-MS analysis. Fortunately, effective strategies for SDS removal do exist, which permits the use of this surfactant in proteomics workflows. Here we highlight an approach that is grounded in the classic technique—protein precipitation. The technique has been updated and has recently seen a revival as a strategy permitting high protein recovery, with exceptional purity. Moreover, with aid of simple disposable spin cartridges, protein precipitation can meet the needs of high throughput, automated, and reproducible proteome purification, enabling the analysis of SDS-containing samples in both top-down and bottom-up formats.

Keywords: sample preparation, mass spectrometry, sodium dodecyl sulfate, GELFrEE, acetone precipitation, ion pairing, protein modification, ProTrap XG

1. Introduction

In proteomics, mass spectrometry (MS) is the tool of choice for comprehensive analytical characterization. Both qualitative identification and quantitative profiling of individual proteins in the mixture are made possible through MS, noting that distinct detection workflows are available to characterize the system. Acknowledging the multiple advances in MS instrumentation as primary contributors to the tremendous growth in proteome characterization, it is essential to realize that MS represents but one component in a complex proteome workflow. Successful proteome characterization ultimately depends on the proper isolation and delivery of purely resolved proteins in high yield to the analytical detection platform. The establishment of consistent "front-end" approaches which can ensure the removal of MS interferences, with unbiased recovery of all protein components, in LC-MS compatible format is an essential first step of the detection platform.

1.1 Top-down versus bottom-up proteomics

Two distinct and complementary workflows exist to characterize proteome samples by mass spectrometry: the bottom-up proteomics (BUP) and top-down proteomics (TDP) methods [1]. BUP is considered the more classical and also most widely used approach to MS-based proteome characterization. In BUP, proteins are first hydrolyzed into smaller peptide segments, typically using an enzyme such as trypsin. Trypsin cleaves at specific amino acid residues, namely on the C-terminal side of lysine and arginine [2], giving rise to predictable peptides possessing a favorable mass range (~1–2 kDa) and charge state (+2 or higher), both of which encourage optimal MS detection. These smaller peptide chains are separated by liquid chromatography (LC), which can be directly coupled to MS by way of electrospray ionization (ESI). Small peptides are readily resolved through reversed phase liquid chromatography. If necessary, they can even be further fractionated by way of orthogonal modes of separation such as ion exchange chromatography [3]. As the peptides elute from the LC column and are directed to the gas phase by ESI, the resulting ions are first profiled by MS to determine the molecular weight of the precursor (intact) peptide molecule. These gas phase molecules are then fragmented, and through a second dimension of mass spectrometry (i.e., tandem MS, or MS/MS), the resulting spectra reveal the mass of the fragments. With aid of computer programs, the tandem MS spectra ultimately carry the mass information to decipher the amino acid sequence of the peptides, which are correlated back to the original protein. BUP offers the potential for deep proteome coverage; using state-of-the-art MS instrumentation, coupled with appropriate separation, it is essentially possible to detect "all" unique proteins of simple proteome systems [4, 5]. Further to this, quantitative analysis is readily possible, using internal [6] or external [7] calibration standards often generated through isotope labelling to correlate MS intensity date with sample concentration.

While BUP is a widespread approach, several researchers have recognized the limitations of the strategy to attain the goals of proteome profiling. In particular, while the detection of multiple peptides stemming from distinct proteins of a proteome mixture does demonstrate the presence of a protein, it is rarely possible to observe "all" peptides stemming from a given protein. In other words, protein sequence coverage is often incomplete. It is possible that important regions of a protein, for example, one containing a post translational modification, may go undetected. Post-translational modifications are but one of the factors leading to greater protein diversity that would be describe by a simple translation of the genome into a proteome. As such, researchers have used the word "proteoform" to describe the distinct chemical entities generated from slight variations in proteins [8]. For example, consider hypothetical protein which might carry 10 possible post-translational modifications. Even if these modifications are binary in nature (i.e., present or not), this would lead to 2^{10} (1024) distinctly modified forms (modforms) for this one protein [9]. Though this simple example may be an over-estimation of protein diversity, it is clear that the proteome carries with it a greater chemical diversity than the genome from which it is derived. Together with dynamic changes in expression, the functional diversity of an organism is only captured through characterization of distinct proteoforms. Given the importance of complete proteome characterization, MS detection workflows should be cognizant of providing data which captures this increased diversity in the sample. BUP methods are generally not suited to profile distinct proteoforms, since the digestion process will not retain the combination of protein modifications that would be retained at the intact molecular level.

By contrast to BUP, with top-down proteomics, proteins are detected by mass spectrometry without subjecting them to enzymatic digestion. Thus, the gas phase protein ions are generally of much higher molecular weight (~10–100 kDa), and also of higher charge state (~+5 to +50). These higher masses demand higher resolution MS platforms, as well as fragmentation approaches suited to generate tandem MS spectra for the intact proteins [10]. Like BUP, the TDP approach also necessitates front-end separation, as well as computational tools to interpret the resulting spectra. Given that the TDP approach begins with the entire protein molecule, it is theoretically possible to localize all possible modifications present on the molecule [11]. In practice, localization of these modifications depends on the quality of the resulting fragmentation spectrum. TDP has grown steadily in its ability to detect an increasing number of proteins and proteoforms in a system. Nonetheless, TDP is still considered to be more technically challenging. Several of these challenges stem from issues related to front-end sample preparation. These include greater challenges related to maintaining protein solubility, both during and after proteome fractionation, together with issues surrounding protein purity which impacts MS sensitivity. In short, while TDP offers greater potential for proteoform characterization, proper sample preparation is even more critical to the success of the approach.

1.2 Protein solubility

The digestion of a protein into peptide fragments offers immediate advantages in terms of easier sample handling. Peptides are generally more soluble than their intact protein counterparts. Certainly, there is likely to be at least one portion of the digested protein that is more soluble and also more readily detected by MS (the proteotypic peptide) [12]. In sharp contrast, the chemical diversity existing at the level of intact proteins is far greater. Proteins vary more greatly in size, charge, and polarity [13]. They also adopt various folding states and therefore can behave unpredictably throughout the various stages of sample preparation.

The solubility of an intact protein is highly influenced by the tertiary structure of the protein. In aqueous systems, proteins adopt a folded structure which shields the more hydrophobic amino acids, while exposing polar or charged residues on their outside surface. This favors the formation of an ordered hydration layer in order to shield the electrostatic regions of the protein; therefore decreasing electrostatic protein-protein interactions. At low ionic strength the shielding of the protein by the hydration layer can be described by the Debye-Huckel theory, which assumes that the protein is surrounded by ions of opposite charge [14]. This causes a reduction in the electrostatic free energy of the protein, resulting in a decrease in activity. This solubility can be described by the equation:

$$ln\left(s_2/s_{2,w}\right) = Z^2\varepsilon^2 N\kappa/[2DRT(1+\kappa a)] \tag{1}$$

where Z is the overall net charge, ε is the electronic charge, N is Avogadro's number, D is the dielectric constant, R is the universal gas constant, T is the temperature, a is the radius of the ionic cloud, I is the ionic strength and κ is given by:

$$\kappa = \sqrt{8\pi N\varepsilon^2/1000DkT}\sqrt{I} \tag{2}$$

$$I = \frac{1}{2}\sum_i C_i Z_i^2 \tag{3}$$

This description of protein solubility only accounts for the electrostatic forces of the protein and not the hydrophobic interactions. Protein solubility in aqueous systems has been further described by two mathematical models, the osmotic second virial coefficient and the preferential binding parameter which are reviewed by Ruckenstein and Shulgin [15].

1.3 SDS in proteomics

Many proteins are poorly soluble in aqueous solution due to the presence of hydrophobic amino acids. Membrane proteins are a class of proteins that are associated with the cell lipid bilayer; these proteins tend to have more hydrophobic regions which can interact more favorably with the lipid membrane than the surround water molecules. Integral (transmembrane) proteins are permanently bound to the membrane, with highly hydrophobic segments of the protein often spanning the membrane itself. Peripheral membrane proteins are temporarily bound to the lipid membrane surface, or to integral membrane proteins. The importance of membrane proteins comes from their role in cell signaling, and regulation of cellular communication. It is estimated that out of all the proteins the in the mammalian genome, 30% are classified as membrane proteins [16]. Membrane proteins represent two thirds of protein targets for potential drugs due to their accessibility [17]. However, these protein targets are often underrepresented in proteomics analysis, owing to reduced solubility.

To improve protein solubility, particularly for membrane proteins, sodium dodecyl sulfate (SDS) may be added [18]. SDS monomers bind to the protein *via* electrostatic and hydrophobic interactions. The hydrophobic binding of SDS to proteins is caused by the relative increase in entropy for interaction of the aliphatic chain of the SDS monomer with hydrophobic regions of the protein, relative to that of interactions between these same respective regions with water molecules. In other words, hydrophobic molecules prefer to come together in aqueous solution. Simultaneously, the electrostatic interactions of SDS with protein are readily recognized; electrostatic attraction between opposing charged ions are a direct result of Coulombic forces. The negatively charged head group of the SDS monomer attracts to cationic side chains of the protein (e.g., deprotonated glutamic and aspartic acid). At SDS concentrations above the critical micelle concentration (CMC), SDS micelles begin to form on the protein around the original monomer. The protein becomes denatured, causing loss of secondary, quaternary and tertiary structure due to replacement of previously favored intra-protein hydrophobic attraction to the more favored SDS-protein interaction [19]. The overall increase in protein solubility is related to the incorporation of the insoluble portions of the protein into the core of the soluble SDS micelle that forms upon these regions. Reynolds and Tanford showed that above 0.5 mM, SDS and protein bind together at a 1.4–1 mass ratio of SDS monomer to protein [20].

Surfactants such as SDS can also be used to disrupt cell membranes which will extract proteins from biologically relevant samples. Lipid membrane disruption and protein solubilization results in the removal of lipid monomers and effective solubilization of membrane proteins [18].

Further to the benefits of membrane extraction and protein solubilization, SDS also has important uses for enhancing protein solubility. Specifically, SDS is a key component of mass-based protein separation by polyacrylamide gel electrophoresis (SDS PAGE). More recently, the introduction of a related technology termed GELFrEE allows for intact protein separation with recovery of fractions sorted by size in the solution phase [21]. Unfortunately, with GELFrEE, samples are collected in the running buffer. Just as in SDS PAGE, the GELFrEE running buffer contains significant quantities of SDS. Therefore, for GELFrEE fractionated proteins to be amenable to LC-MS analysis, the samples must first be purified to remove the surfactant to levels that no longer interfere with ESI-MS sensitivity. A maximal level of 100 ppm SDS is often quoted to permit LC-MS analysis [22]. However, optimal MS sensitivity is only achieved when SDS is reduced to ~10 ppm or less [23]. For a sample initially containing 0.1% SDS, this would constitute >99% removal of SDS from the sample.

2. Protein purification strategies

A number of methods have been reported for the removal of SDS from proteomics samples, both at the peptide and at the intact protein level. These include classical dialysis [20], ultrafiltration [24], solid phase extraction [25], electrophoretic approaches [23], and protein precipitation [22]. These methods vary greatly in terms of their expected protein yield, level of purity, as well as their throughput and relative ease of use. The ideal SDS removal technique would provide *quantitative* recovery, with *completely* SDS removal from protein or peptide samples. While such a technique does not currently exist, some techniques have proven to be very close.

A promising approach to SDS removal was proposed by Wiśniewski et al., being described initially as a "universal" protocol for protein sample preparation [24]. The technique, known as filter aided sample preparation (FASP), relies on molecular weight cut off (MWCO) filters (~3 to 10 kDa in porosity) to selectively remove low molecular weight containments including SDS, while retaining the high mass proteins. Following purification by filtration, proteins are enzymatically digested on the filter to liberate peptides which are collected for MS analysis. While the technique is generally very effective at depleting SDS to levels permitting MS analysis, several researchers have noted the variable, if not poor recovery of proteins when processed by FASP, which may fall below 50% [26–29]. Protein loss can be attributed to the poor retention of low-mass proteins (<5 kDa), peptide fragments that do not readily pass through the filter (>5 kDa), or protein interactions with the filter giving rise to poor protein/peptide solubility. Nonetheless, FASP remains a favored approach to handle SDS-containing samples, owing perhaps to the ease of using disposable filter cartridges which generally provide positive proteome results. Perhaps the popularity of this technique also falls to a perceived lack of favorable alternative.

Consider dialysis as an alternative SDS depletion strategy; it should be recognized that this classic protein purification approach is, in fact, ineffective for the purpose of SDS depletion ahead of LC-MS analysis. With conventional dialysis, low molecular weight components may be eliminated by passive diffusion across a molecular weight cut-off membrane. Given an appropriate concentration gradient, and sufficient time one would assume the technique could eventually deplete SDS from a sample. However, the tight binding interaction between SDS and protein implies a significant portion of the SDS will remain complexed to the larger protein molecules. These bound surfactants will not be eliminated, even with exhaustive dialysis. For optimal MS analysis, both free and protein-bound SDS must be removed. Therefore, to fully deplete SDS from the sample, the interaction between SDS and protein must be overcome. FASP facilitates this depletion by adding compounds which weaken the SDS-protein interactions (e.g., urea, sodium deoxycholate) [30].

Similar to dialysis, several chromatographic approaches only offer partial depletion of SDS. Size exclusion chromatography for example, exploits the size difference between SDS and protein, but would not intrinsically separate SDS-protein complexes. Similarly, ion exchange chromatography can be used to selectively capture negatively charged SDS (strong anion exchange) [31], or positively charged proteins (strong cation exchange) [25], while allowing the opposing compound to flush through. While recognizing that multiple chromatographic approaches have been successfully incorporated into detergent-based workflows, it is noted that the existence of SDS-protein complexes implies that residual SDS is likely, and thus MS analysis is below optimal performance.

3. Protein precipitation

As with all purification strategies, maintain high protein recovery is of utmost importance. Biased sample loss can lead to incomplete proteome characterization, or may skew quantitative analysis by mass spectrometry. With this in mind, consider protein precipitation as a potential means for detergent purification ahead of LC-MS. Precipitation is recognized as a classic approach, predating chromatography and even electrophoresis. It has been used to isolate specific proteins, generally through the addition of a precipitating agent which acts to lower the solubility of the protein. Common precipitating agents include salts [32], organic solvents [22], acids [33], or polymers [34]. Heat and mechanical agitation may also lead to precipitation. In all cases, precipitating agents act to alter the interaction between protein and the solvent system (typically water). Often this occurs in parallel with modifying the tertiary structure of the protein. Once the proteins are made insoluble, they will aggregate and can be isolated from the supernatant solvent, which still contains the interfering contaminants.

Early scientific reports of protein precipitation focused on the isolation of specific proteins from sources including milk or plasma. The classic work by Hofmeister and his student Lewis in 1888 [35], measured the ability of a variety of salts to precipitate proteins from egg albumin and other sources. It was found that the precipitating ability of a salt was dependent on (1) the type of salt/precipitating agent used, (2) the type of protein, and (3) the concentration of protein. These factors seem to play a role in other forms of protein precipitation. Today, ammonium sulfate is still commonly used to precipitate and concentrate proteins from solution. An advantage of salt precipitation is its ability to selectively precipitate proteins from a solution, thus lending a form of separation. For example, Jiang et al. applied ammonium sulfate precipitation to deplete the highly abundant albumin protein from plasma [36]. In their report, 30% NH_4SO_4 was used to precipitate proteins from plasma while leaving albumin in solution.

By contrast to salt, organic solvents can be used to induce precipitation of proteins. Precipitation has been seen through the use of a verity of water miscible solvents such as ethanol, acetonitrile, methanol, and acetone. Focusing on acetone precipitation, the solvent was first reported a century ago as a method to remove water from blood in an attempt to obtain a possible blood alternative [37]. Since then acetone has been applied in a number of applications including protein concentration, metabolite isolation and SDS detergent removal. Protein precipitation as a whole is a very common sample preparation strategy; in 1990 it was estimated that 80% of all proteomic experiments contained a protein precipitation step [38]. However, focusing on proteome analysis with MS, the popularity of protein precipitation did not keep pace with alternative strategies including chromatographic approaches. This is most likely due to two factors: (1) the *apparent* variable protein recovery, as reported across different labs and for different proteins, (2) a *perceived* difficulty in isolating the protein pellet through manual pipetting. As is described below, both of these shortcomings of solvent precipitation have now been overcome.

3.1 Acetone precipitation mechanism

Protein solubility is a physiochemical property which describes the amount of protein that can be solvated and therefore dissolved by a given solvent. As has been already described, sample additives such as SDS can influence protein solubility, as can protein specific factors include the amino acid sequence, molecular weight, and protein conformation. Altering the solvent conditions will have an influence on protein solubility, noting particularly the solution pH, ionic strength, temperature, and solvent polarity. The addition of organic solvent to a water soluble protein generally lends a reduction in protein solubility, leading to protein aggregation. This ultimately is caused by increasing the protein-protein interactions (electrostatic or hydrophobic attractions) while decreasing the solvation ability of the solvent.

As a starting point, consider that it is possible to induce selective protein precipitation by using a modestly low concentration of organic solvent (if not a purely aqueous system) and controlling the pH of the solution. Proteins are known to have reduced solubility at their intrinsic isoelectric point (pI) due to the decrease of electrostatic charge on the surface of the protein; this decrease charge results in a decrease in interaction between water and the macromolecule and protein-protein interactions dominated. Although proteins have a reduced solubility at their pI, they are not necessarily insoluble. Through the addition of low concentrations of organic solvent the reduced solubility can be further reduced causing precipitation. This allows for the selective removal of proteins according to isoelectric point. This is the basis of a technique known as Cohn fractionation of blood plasma, where proteins such as human serum albumin, serum gamma globulin, fibrinogen, thrombin, and a few others are isolated from plasma [39].

Precipitation induced through the addition of organic solvent is thought to be caused by the decrease in the dielectric constant of the solution (water has a dielectric ~70 while most organic solvents are ~20). As the amount of organic solvent is increased, the solvating power of water will decrease. It has been shown that ethanol binds water more strongly than it does proteins [40]. This is thought to extend to other organic solvents. The decrease in solvation causes the hydration sphere around the proteins to shrink. The overall shielding of charged region is therefore decreased. Following this, it is thought that the increased electrostatic interaction between *opposing* charged regions on distinct protein molecules will cause the proteins to aggregate. As described below:

$$|F| = k_e \frac{|q_1 q_2|}{r^2} \tag{4}$$

$$k_e = 1/(4\pi\varepsilon_0\varepsilon) \tag{5}$$

F is the magnitude of the force between the charges (for like charges, this is a repulsive force; opposing charged species attract), q_i is the magnitude of charge, r is the distance between the two charges, ε_o is the permittivity of free space, and ε is the relative permittivity of the solution. As can be seen, the dielectric constant is in the denominator of the equation, meaning that the magnitude of the Coulombic force increases as the dielectric decreases [41, 14]. Water has a high dielectric, while organic solvents have lower dielectrics. Therefore, charged species are more strongly attracted in organic solvent. Also, it is noted that electrostatic interactions are not the only force to consider. Dipolar van der Walls forces are thought to play an important role as well in bringing proteins together. However, as presented below, there are issues with the underlying assumption of this model, which bring to question the validity of this model.

The addition of acetone to protein samples is generally performed at low temperature. This is done to prevent protein denaturation which has been found to occur quickly above 10°C [41]. The denaturation effect of organic solvent is due to the interaction of the more favorable interaction of the hydrophobic regions of the protein compared to water. This decreases the entropic loss which occurs when the protein unfolds, promoting denaturation. The use of cold temperatures is thought to reduce the conformational flexibility of proteins preventing the organic solvent from accessing the hidden hydrophobic regions. Acetone has been found to be less denaturing than ethanol for protein precipitation.

3.2 The role of salt in acetone precipitation

The current theory of solvent precipitation described above is limited in its ability to explain a more recent finding related to organic solvent precipitation. Specifically, experimental evidence has brought to question the validity of electrostatic attraction *between proteins* as a fundamental premise of protein precipitation in organic solvent.

Work in our lab has shown that for a series of protein standards, as well as for complex protein mixtures, the addition of 80% acetone does not in itself induce the precipitation of proteins [42]. In other words, it has been shown that all proteins are in fact completely soluble in 80% acetone! Such a finding is certainly surprising, as 80% acetone. However, upon addition of even minimal amounts of salt (sodium chloride, or other ionic species), these proteins will immediately precipitate from the organic solvent system [42]. **Figure 1** depicts this trend for the protein bovine serum albumin (BSA) precipitated in the presence of increasing concentrations of SDS.

While it may be easy to assume that the phenomenon is related to the solubilizing or denaturing properties of SDS, this assumption does not explain why the absence of SDS prevents the protein from precipitating. In fact, it was discovered that it was the ionic character of SDS that gave rise to this trend. Identical plots can be produced using NaCl in the precipitating solvent system; only above a threshold level will the protein begin to precipitate.

In this study, it was shown that the amount of salt required to induce quantitative protein precipitation was dependent on (1) the concentration of acetone in the solvent system, (2) the type of protein in the sample, (3) the concentration of protein in the sample [42].

To explain why traces of salt are needed to induce near-quantitative precipitation of proteins in acetone, we propose a theory of ion pairing in organic solvent. This theory is only a slight perturbation of the prevailing theory that protein aggregation is related to electrostatic attraction. The primary difference simply relates to the identity of the species being attracted.

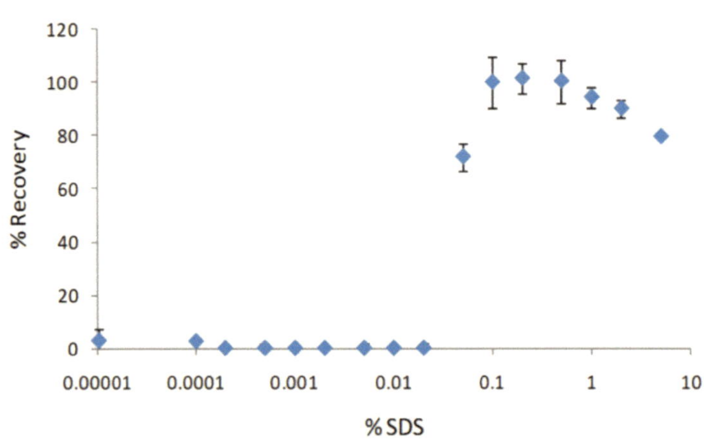

Figure 1.
The recovery of bovine serum albumin (initially 1 g/L) following precipitation in 80% acetone with inclusion of increasing concentrations of SDS in the initial solution. When low concentrations of SDS are present, poor recovery is obtained, implying that the protein is still soluble in 80% acetone.

In aqueous systems, ionic species exist in solution as hydrated spheres, the water acting to partially shield the charge of the ion. This shielding prevents opposing charged ions to attract (pair) as they otherwise would. This is essentially the premise of Coulomb's law. Likewise, it is understood that proteins also carry surface charge (both positive and negative charges can exist simultaneously). As has already been described when considering protein solubility at its isoelectric point, the primary reason for protein solubility in water relates to electrostatic repulsions. This is of course why proteins are least soluble at a pH equal to their respective isoelectric point.

Suppose now that a protein is placed in a solvent system of lower dielectric strength, such as acetone. The hydration spheres surrounding the charges will be reduced, allowing the exposed ions of opposing charge to experience a higher attractive force—according to Coulombs law. What is essential to realize is that it is the attraction between protein and salt that are of interest during solvent precipitation. Whereas in water, the salt ions are hydrated, in organic solvent the lower dielectric allows these ions to pair with opposing charged residues on the proteins [43]. The net result is to effectively neutralize the surface charge of the protein. Adding salt in organic solvent essentially performs the same task as titrating the solution pH to the protein's isoelectric point. Once the protein's surface charge is neutralized, the repulsive electrostatic forces between proteins are minimized. This allows van der Waals forces to take over, allowing the hydrophobic portions of proteins to aggregate. Precipitation is therefore intrinsically connected to electrostatic effects. However, these specific interactions are also tied to the presence of ionic species (salts) in the system.

3.3 Protein recovery and purity from acetone precipitation

Throughout the literature, variable protein recovery has been reported through acetone precipitation. Certainly, in light of the more recent findings that salt is required for protein recovery in acetone, low yield may potentially be explained by a lack of understanding of the variables influencing protein recovery. It has been suggested that recovery through acetone precipitation is protein specific, and also depends on sample concentration, as well as the presence of sample additives (beyond salt). Reported recoveries from acetone precipitation have ranged from extremely low to near quantitative.

Thongboonkerd et al. measured the protein recovery from urine using acetone precipitation [44]. Increasing the percentage of acetone (from 10 to 90%) improved protein recovery, though only 40% yield was obtained in 90% acetone. Barritault et al. found acetone precipitation provided high (>95%) in a short period of time for high concentrations of samples but required a far greater time (overnight) to provide similar yield to more dilute samples [45].

Srivastava and Srivastava employed 50% acetone to enrich gamma-crystallin from human eye lenses [46]. The hydrophobic, low molecular weight (20 kDa) protein remains soluble in a 50% acetone solution. Ashri et al. examined the effect of acetone concentration on removing proteins from plasma and report that only 70% acetone is required to induce near quantitative precipitation of proteins [47]. As concentration was decreased, yield too decreased. Lin et al. determined that an 85% acetone solution produced optimal precipitation, however it is noted that 80% was only marginally lower in efficiency [48].

Low recovery ranging from 40 to 50% was obtained by Sickmann et al. when purifying human cerebrospinal fluid [29]. However, Yuan et al. found recoveries of 94% when applying acetone to a similar sample of human cerebrospinal fluid [49]. No significant difference in protocol was reported, implying that recovery differences are either related to an unreported difference, or are perhaps related to accidental sample loss contributed through manual sample pipetting.

Using the newly established principle of including sufficient quantities of salt (10–100 mM) to ensure maximal protein recovery, our group has consistently reported protein recovery above 90%. Often times, recovery is statistically considered to be quantitative (>99%). Under appropriate conditions, high protein recovery is obtained for all sample types (soluble/hydrophobic, high/low molecular weight) and over a range of protein concentrations (sub microgram to high milligram per milliliter starting amounts). Furthermore, the ability of acetone precipitation to eliminate SDS has been examined by Botelho et al. [22]. In this report, a sample initially containing 2% SDS can be depleted below 0.005% with inclusion of an additional washing step to rinse the protein pellet. Still, to obtain high recovery and purity, the pipetting step must be performed with great precision.

3.4 Issues with protein precipitation

While a so-called "universal" strategy for front-end sample preparation does not exist, it can be suggested that acetone precipitation presents a powerful starting point. Despite the potential for high yield and purity, a number of potential disadvantages exist. Here we focus attention on the potential of acetone to induce chemical modifications in the sample, as well as difficulties in pipetting the sample.

3.4.1 Protein modification by acetone

Simpson et al. described a +40 Da peptide modification induced by the presence of acetone in solution [50]. They correlate this modification to peptide sequences which contain a glycine residue as the second amino acid in the sequence. They found modifications to exist within 1 hour of reaction and a rate constant of $0.29 \pm 0.01\ h^{-1}$ for a number of different peptides. There was no evidence to suggest that acetone can modify at the protein level. However this is still an issue in bottom-up proteomics as any unknown peptide modification would skew results through the splitting of signal intensity leading to incorrect conclusions. The complete removal of acetone after precipitation is required to prevent this modification, therefore proper steps should be taken to guarantee complete removal such as allowing significant time for the pellets to air dry, or drying under vacuum.

In an independent study, it was shown that acetone may induce modifications at the intact protein level, manifesting as a +98 u mass shift in the mass spectrum of acetone-precipitated proteins [51]. The +98 u artifact was speculated to originate from the aldol condensation of acetone to form diacetone alcohol and mesityl oxide, which in turn reacts by nucleophilic attack. The degree of protein modification was temperature and time sensitive, wherein over 90% of cytochrome c was detected in the modified form following 1 hour incubation in acetone at 0°C. However, it was also shown that this modification was highly dependent on solution pH. By precipitated the protein in an acidic environment, any possibility of modifying the protein was eliminated.

3.4.2 Technical difficulties of protein precipitation

The second disadvantage of acetone precipitation over competing sample preparation techniques such as filtration or chromatography involves the technical difficulty in separating the protein pellet from the organic solvent. If one removes too little solvent, contaminants will remain in the protein sample which may interfere with subsequent analysis. However if too much supernatant is removed, disruption and loss of the protein pellet can occur which will cause reduce protein yield. This is especially the case when precipitating dilute protein samples; here the protein pellet may not even be visible to the naked eye. Protein losses due to accidental pipetting of the pellet are almost unavoidable, without considerable care, and a strong familiarity of the technique. Botelho et al. suggest the use of a wash step to allow one to leave behind larger portions of acetone; however each wash implies another manipulation which takes time, and introduces the possibility of error [22]. Improper formation of the protein pellet after the wash may also cause protein loss. The pellet is not necessarily strongly bound to itself, or to the vial surface. Regardless of the inherent ability of acetone precipitation to aggregate protein, the method is not useful without a simple means of isolating the pellet.

3.5 The ProTrap XG

Filtration cartridges are commonly used in multiple applications such as DNA isolation, trapping cells, and of course for protein isolation. A number of microcentrifuge filtration devices are currently available on the market. However, these devices are not inherently designed to recover precipitated precipitation. With precipitation, a certain period of time is required to allow the protein to aggregate. Thus, filtration should not be initiated until after the pellet has properly formed. However, once the protein has aggregated, the macroscopic structure should enable filtration as a reliable means of isolating the pellet. Following precipitation, the protein pellet must now be resolubilized, which again requires the absence of filtration during this stage.

The ProTrap XG is a disposable spin cartridge used to recover precipitated protein. As shown in **Figure 2**, the cartridge includes a detachable plug at the base of a Teflon membrane filter. Compared to molecular weight cutoff cartridges (seen in dialysis or FASP), the ProTrap XG membrane is of relatively large porosity, allowing rapid flow of solvent through the cartridge. However, the membrane is still sufficient to recover aggregated protein in high yield. As demonstrated, the ProTrap XG allows near quantitative recovery of proteins (**Figure 3**), including sub microgram levels [52].

The ProTrap XG also includes a solid phase extraction cartridge which can be attached to the base of the filter. Such cartridge enables downstream cleanup or separation of proteins or peptides following the precipitation step.

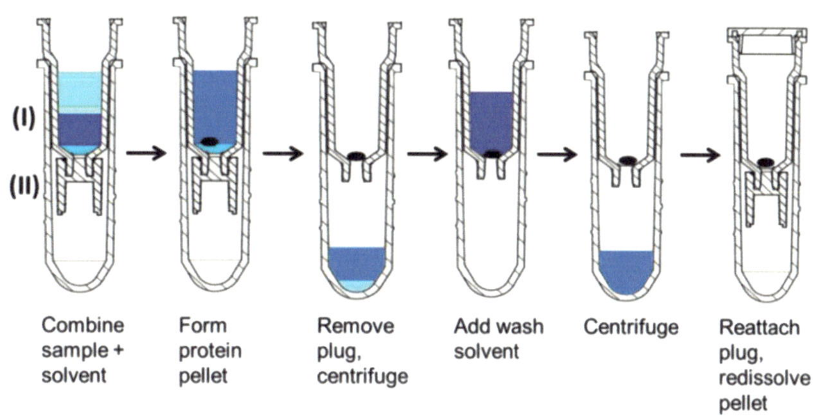

Figure 2.
The ProTrap XG is a two-stage cartridge combining an upper membrane filtration cartridge (I) with a detachable plug (II) to capture precipitated proteins. Acetone precipitation of proteins with the ProTrap XG allows for recovery in high yield and with high purity.

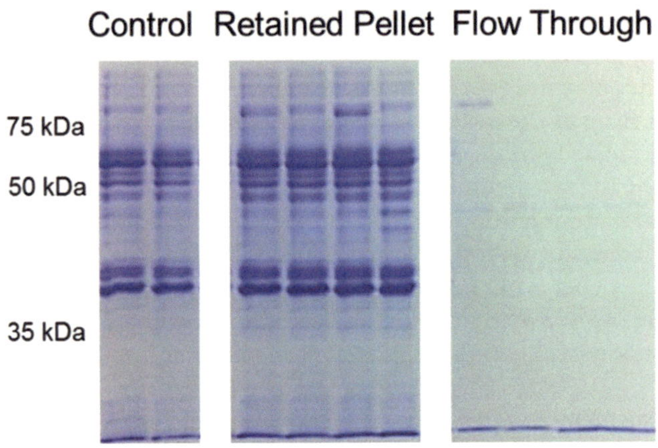

Figure 3.
Acetone precipitation of a proteome mixture using the ProTrap XG demonstrated high recovery of the aggregated protein above the membrane filter.

3.6 How to resolubilize proteins after precipitation

Perhaps the main disadvantage of protein precipitation is that the resulting solid protein pellet is not directly amenable to MS analysis. Specifically, the protein must first be resolubilized prior to further analysis. Although a number of strategies exist, currently there are no "perfect methods" to allow for resolubilization of the protein pellet. Any solubilization method must be compatible with downstream analysis. Naturally, this means that SDS is not available as a solubilizing additive. The simplest approach is to simply use water, or more specifically the solvent system used for LC-MS analysis (typically 5% acetonitrile, 0.1% formic acid in water). The addition of 5% acetonitrile does little to aid protein dissolution, though pH is an important addition.

A number of MS-compatible "cleavable" surfactants have been proposed to resolubilize proteins [53]. But they are generally quite expensive. High concentrations of urea (8 M) are common to resolubilize proteins. This additive is readily removed through reversed phase, so it can be considered compatible with LC-MS. Also, the addition of digestive enzymes including trypsin will aid in the dissolution of proteins, by liberating the more soluble peptide counterparts into solution. As a disadvantage, such an approach is only suited to bottom-up proteomics.

An effective strategy to dissolve intact proteins is to use high concentrations of the organic acid, including formic acid (50–80% acid by volume). This solvent system is shown to dissolve proteins as effectively as high concentrations of SDS [54]. A concern with formic acid is its propensity to rapidly modify the protein, causing the addition of +28 u adducts in the resulting mass spectrum (28 = CO modification). However, this reaction can be reduced, if not entirely eliminated by maintaining a low temperature during incubation of the protein in concentrated formic acid [55]. At −20°C, no protein modifications are detected. At the same time, the reduced temperature does not deter proteome resolubilization. Finally, formic acid is compatible with LC-MS analysis, as the proteins are retained on the reversed phase column while the acid flushes through.

4. Conclusions

Proteome analysis by LC-MS demands isolation of proteins in high purity, particularly when the sample is contaminated with known MS interferences such as SDS. While several options are available for protein purification, only some are suited to remove the SDS that is tightly associated with protein. Solvent precipitation with high concentrations of acetone shows promise as a high recovery, high purity approach for proteome sample preparation. The requirement to include ionic species (salt) in the precipitating solvent lends knew knowledge to the underlying mechanism controlling solvent precipitation. So long as the supernatant can be removed without disturbing the protein pellet, exceptional high recovery and purity can be expected. This is facilitated by filtering the sample using disposable spin cartridges. The resulting pellet can be resolubilized in MS compatible solvents such as concentrated formic acid. This workflow presents an effective approach to detergent-based proteome analysis in both top-down and bottom-up formats. As such, acetone precipitation should see increased use as researchers exploit the advantages of SDS for protein extraction and separation.

Acknowledgements

Work cited in this chapter by the author has been funded by a Discovery Grant from the National Sciences and Engineering Research Council of Canada (NSERC).

Conflict of interest

A.D. holds a patent on the ProTrap XG, and is affiliated with Proteoform Scientific, the company who holds license to commercialize and distribute the ProTrap XG. All results presented in this chapter have been verified through independent peer review and the original scientific articles have been cited accordingly.

Author details

Alan Doucette[1*] and Andrew Crowell[2]

1 Department of Chemistry, Dalhousie University, Halifax, Canada

2 Verschuren Center, Sydney, Canada

*Address all correspondence to: alan.doucette@dal.ca

References

[1] Resing KA, Ahn NG. Proteomics strategies for protein identification. FEBS Letters. 2005;**579**:885-889. DOI: 10.1016/J.FEBSLET.2004.12.001

[2] Olsen JV, Ong S-E, Mann M. Trypsin cleaves exclusively C-terminal to arginine and lysine residues. Molecular & Cellular Proteomics. 2004;**3**:608-614. DOI: 10.1074/mcp.T400003-MCP200

[3] Washburn MP, Wolters D, Yates JR. Large-scale analysis of the yeast proteome by multidimensional protein identification technology. Nature Biotechnology. 2001;**19**:242-247. DOI: 10.1038/85686

[4] Hebert AS, Richards AL, Bailey DJ, et al. The one hour yeast proteome. Molecular & Cellular Proteomics. 2014;**13**:339-347. DOI: 10.1074/mcp.M113.034769

[5] Picotti P, Clément-Ziza M, Lam H, et al. A complete mass-spectrometric map of the yeast proteome applied to quantitative trait analysis. Nature. 2013;**494**:266-270. DOI: 10.1038/nature11835

[6] Ong S-E, Blagoev B, Kratchmarova I, et al. Stable isotope labeling by amino acids in cell culture, SILAC, as a simple and accurate approach to expression proteomics. Molecular & Cellular Proteomics. 2002;**1**:376-386. DOI: 10.1074/mcp.M200025-MCP200

[7] Oda Y, Huang K, Cross FR, et al. Accurate quantitation of protein expression and site-specific phosphorylation. Proceedings of the National Academy of Sciences of the United States of America. 1999;**96**: 6591-6596. DOI: 10.1073/pnas.96.12.6591

[8] Smith LM, Kelleher NL, Proteomics TC for TD, et al. Proteoform: A single term describing protein complexity. Nature Methods. 2013;**10**:186-187. DOI: 10.1038/nmeth.2369

[9] Compton PD, Kelleher NL, Gunawardena J. Estimating the distribution of protein post-translational modification states by mass spectrometry. Journal of Proteome Research. 2018;**17**:2727-2734. DOI: 10.1021/acs.jproteome.8b00150

[10] Anderson LC, DeHart CJ, Kaiser NK, et al. Identification and characterization of human proteoforms by top-down LC-21 Tesla FT-ICR mass spectrometry. Journal of Proteome Research. 2017;**16**:1087-1096. DOI: 10.1021/acs.jproteome.6b00696

[11] Siuti N, Kelleher NL. Decoding protein modifications using top-down mass spectrometry. Nature Methods. 2007;**4**:817-821. DOI: 10.1038/nmeth1097

[12] Kuster B, Schirle M, Mallick P, et al. Scoring proteomes with proteotypic peptide probes. Nature Reviews. Molecular Cell Biology. 2005;**6**:577-583. DOI: 10.1038/nrm1683

[13] Doucette AA, Tran JC, Wall MJ, et al. Intact proteome fractionation strategies compatible with mass spectrometry. Expert Review of Proteomics. 2011;**8**:787-800. DOI: 10.1586/epr.11.67

[14] Arakawa T, Timasheff SN. Theory of protein solubility. Methods in Enzymology. 1985;**114**:49-77. DOI: 10.1016/0076-6879(85)14005-X

[15] Ruckenstein E, Shulgin IL. Effect of salts and organic additives on the solubility of proteins in aqueous solutions. Advances in Colloid and Interface Science. 2006;**123-126**:97-103. DOI: 10.1016/J.CIS.2006.05.018

[16] Almén M, Nordström KJ, Fredriksson R, et al. Mapping the human membrane proteome: A majority of the human membrane proteins can

be classified according to function and evolutionary origin. BMC Biology. 2009;7:50. DOI: 10.1186/1741-7007-7-50

[17] Kohnke PL, Mulligan SP, Christopherson RI. Membrane proteomics for leukemia classification and drug target identification. Current Opinion in Molecular Therapeutics. 2009;**11**:603-610

[18] le Maire M, Champeil P, Moller JV. Interaction of membrane proteins and lipids with solubilizing detergents. Biochimica et Biophysica Acta. 2000;**1508**:86-111. DOI: 10.1016/S0304-4157(00)00010-1

[19] Bhuyan AK. On the mechanism of SDS-induced protein denaturation. Biopolymers. 2010;**93**:186-199. DOI: 10.1002/bip.21318

[20] Reynolds JA, Tanford C. Binding of dodecyl sulfate to proteins at high binding ratios. Possible implications for the state of proteins in biological membranes. Proceedings of the National Academy of Sciences of the United States of America. 1970;**66**:1002-1007. DOI: 10.1073/pnas.66.3.1002

[21] Tran JC, Doucette AA. Gel-eluted liquid fraction entrapment electrophoresis: An electrophoretic method for broad molecular weight range proteome separation. Analytical Chemistry. 2008;**80**:1568-1573. DOI: 10.1021/ac702197w

[22] Botelho D, Wall MJ, Vieira DB, et al. Top-down and bottom-up proteomics of SDS-containing solutions following mass-based separation. Journal of Proteome Research. 2010;**9**:2863-2870. DOI: 10.1021/pr900949p

[23] Kachuk C, Faulkner M, Liu F, et al. Automated SDS depletion for mass spectrometry of intact membrane proteins though transmembrane electrophoresis. Journal of Proteome Research. 2016;**15**:2634-2642. DOI: 10.1021/acs.jproteome.6b00199

[24] Wiśniewski JR, Zougman A, Nagaraj N, et al. Universal sample preparation method for proteome analysis. Nature Methods. 2009;**6**:359-362. DOI: 10.1038/nmeth.1322

[25] Sun D, Wang N, Li L. Integrated SDS removal and peptide separation by strong-cation exchange liquid chromatography for SDS-assisted shotgun proteome analysis. Journal of Proteome Research. 2012;**11**:818-828. DOI: 10.1021/pr200676v

[26] Wiśniewski JR, Zielinska DF, Mann M. Comparison of ultrafiltration units for proteomic and N-glycoproteomic analysis by the filter-aided sample preparation method. Analytical Biochemistry. 2011;**410**:307-309. DOI: 10.1016/j.ab.2010.12.004

[27] Kachuk C, Stephen K, Doucette A. Comparison of sodium dodecyl sulfate depletion techniques for proteome analysis by mass spectrometry. Journal of Chromatography. A. 2015;**1418**:158-166. DOI: 10.1016/j.chroma.2015.09.042

[28] Manza LL, Stamer SL, Ham A-JL, et al. Sample preparation and digestion for proteomic analyses using spin filters. Proteomics. 2005;**5**:1742-1745. DOI: 10.1002/pmic.200401063

[29] Sickmann A, Dormeyer W, Wortelkamp S, et al. Identification of proteins from human cerebrospinal fluid, separated by two-dimensional polyacrylamide gel electrophoresis. Electrophoresis. 2000;**21**:2721-2728. DOI: 13<2721::AID-ELPS2721>3.0.CO;2-3

[30] Erde J, Loo RRO, Loo JA. Enhanced FASP (eFASP) to increase proteome coverage and sample recovery for quantitative proteomic experiments.

Journal of Proteome Research. 2014;**13**:1885-1895. DOI: 10.1021/pr4010019

[31] Vissers JPC, Chervet J-P, Salzmann J-P. Sodium dodecyl sulphate removal from tryptic digest samples for on-line capillary liquid chromatography/electrospray mass spectrometry. Journal of Mass Spectrometry. 1996;**31**:1021-1027. DOI: 10.1002/(SICI)1096-9888(199609)31:9<1021::AID-JMS384>3.0.CO;2-G

[32] Chick H, Martin CJ. The precipitation of egg-albumin by ammonium sulphate. A contribution to the theory of the "salting-out" of proteins. The Biochemical Journal. 1913;7:380-398

[33] Sivaraman T, Kumar TKS, Jayaraman G, et al. The mechanism of 2,2,2-trichloroacetic acid-induced protein precipitation. Journal of Protein Chemistry. 1997;**16**:291-297. DOI: 10.1023/A:1026357009886

[34] Zeppezauer M, Brishammar S. Protein precipitation by uncharged water-soluble polymers. Biochimica et Biophysica Acta (BBA): Biophysics including Photosynthesis. 1965;**94**:581-583. DOI: 10.1016/0926-6585(65)90069-5

[35] Kunz W, Henle J, Ninham BW. 'Zur Lehre von der Wirkung der Salze' (about the science of the effect of salts): Franz Hofmeister's historical papers. Current Opinion in Colloid & Interface Science. 2004;**9**:19-37. DOI: 10.1016/j.cocis.2004.05.005

[36] Jiang L, He L, Fountoulakis M. Comparison of protein precipitation methods for sample preparation prior to proteomic analysis. Journal of Chromatography. A. 2004;**1023**:317-320. DOI: 10.1016/J.CHROMA.2003.10.029

[37] Piettre VA. On the separation of proteins of the serum. Comptes Rendus de l'Académie des Sciences. 1920;**170**:1466-1468

[38] Englard S, Seifter S. Precipitation techniques. Methods in Enzymology. 1990;**182**:285-300. DOI: 10.1016/0076-6879(90)82024-V

[39] Cohn EJ, Gurd FRN, Surgenor DM, et al. A system for the separation of the components of human blood: Quantitative procedures for the separation of the protein components of human plasma. Journal of the American Chemical Society. 1950;**72**:465-474. DOI: 10.1021/ja01157a122

[40] van Oss CJ. On the mechanism of the cold ethanol precipitation method of plasma protein fractionation. Journal of Protein Chemistry. 1989;**8**:661-668. DOI: 10.1007/BF01025606

[41] Scopes RK. Protein purification. In: Principles and Practice. New York: Springer-Verlag; 1994. DOI: 10.1007/978-1-4757-2333-5

[42] Crowell AMJ, Wall MJ, Doucette AA. Maximizing recovery of water-soluble proteins through acetone precipitation. Analytica Chimica Acta. 2013;**796**:48-54. DOI: 10.1016/j.aca.2013.08.005

[43] Marcus Y, Hefter G. Ion pairing. Chemical Reviews. 2006;**106**:4585-4621. DOI: 10.1021/CR040087X

[44] Thongboonkerd V, Chutipongtanate S, Kanlaya R. Systematic evaluation of sample preparation methods for gel-based human urinary proteomics: Quantity, quality, and variability. Journal of Proteome Research. 2006;**5**:183-191. DOI: 10.1021/pr0502525

[45] Barritault D, Expert-Bezancon A, Guérin MF, et al. The use of acetone precipitation in the isolation of ribosomal proteins. European Journal of Biochemistry. 1976;**63**:131-135. DOI: 10.1111/j.1432-1033.1976.tb10215.x

[46] Srivastava OP, Srivastava K. Purification of gamma-crystallin from human lenses by acetone precipitation method. Current Eye Research.

1998;**17**:1074-1081. DOI: 10.1076/ceyr.17.11.1074.5228

[47] Ashri NY, Abdel-Rehim M. Sample treatment based on extraction techniques in biological matrices. Bioanalysis. 2011;**3**:2003-2018. DOI: 10.4155/bio.11.201

[48] Lin Y, Liu H, Liu Z, et al. Shotgun analysis of membrane proteomes using a novel combinative strategy of solution-based sample preparation coupled with liquid chromatography–tandem mass spectrometry. Journal of Chromatography B. 2012;**901**:18-24. DOI: 10.1016/j.jchromb.2012.05.035

[49] Yuan X, Russell T, Wood G, et al. Analysis of the human lumbar cerebrospinal fluid proteome. Electrophoresis. 2002;**23**:1185-1196. DOI: 10.1002/1522-2683(200204)23:7/8<1185::AID-ELPS1185>3.0.CO;2-G

[50] Simpson DM, Beynon RJ. Acetone precipitation of proteins and the modification of peptides. Journal of Proteome Research. 2010;**9**:444-450. DOI: 10.1021/pr900806x

[51] Güray MZ, Zheng S, Doucette AA. Mass spectrometry of intact proteins reveals +98 u chemical artifacts following precipitation in acetone. Journal of Proteome Research. 2017;**16**:889-897. DOI: 10.1021/acs.jproteome.6b00841

[52] Crowell AMJ, MacLellan DL, Doucette AA. A two-stage spin cartridge for integrated protein precipitation, digestion and SDS removal in a comparative bottom-up proteomics workflow. Journal of Proteomics. 2015;**118**:140-150. DOI: 10.1016/j.jprot.2014.09.030

[53] Hellberg P-E, Bergström K, Holmberg K. Cleavable surfactants. Journal of Surfactants and Detergents. 2000;**3**:81-91. DOI: 10.1007/s11743-000-0118-z

[54] Doucette AA, Vieira DB, Orton DJ, et al. Resolubilization of precipitated intact membrane proteins with cold formic acid for analysis by mass spectrometry. Journal of Proteome Research. 2014;**13**:6001-6012. DOI: 10.1021/pr500864a

[55] Zheng S, Doucette AA. Preventing N- and O-formylation of proteins when incubated in concentrated formic acid. Proteomics. 2016;**16**:1059-1068. DOI: 10.1002/pmic.201500366

Chapter 3

Purification of Proteins: Between Meaning and Different Methods

Hanaa E.A. Amer

Abstract

Purification of protein attracts the scientists' attention toward science in 1926, as Somnar started purification and crystallization of urease from yeast. As years go forward, protein purification strategies updated from using dextrose passing through DEAE-cellulose and ended by affinity chromatography. In this chapter we will describe some of the concepts and the differences between traditional and novel purification methods, in order to point out how a researcher can start the protein purification techniques. Different fundamental purification steps from plant sample will be discussed in this chapter including isolation, concentration, ion exchanger, and affinity chromatography, as well as the important additives that a researcher should add in order to gain a high purification fold.

Keywords: protein, purification, ion exchanger, gel filtration, affinity chromatography

1. Introduction

Purification is an intrinsic objective for profound and exact portrayal for an objective protein, a target protein. As scientists assume that proteins are responsible for bad and good roles throughout everyday life, in this manner we ought to apply more endeavors to definitely comprehend proteins' structure and characters. This objective cannot be achieved else we acquired a highly purified protein which is unfortunately usually present in crowed with different protein communities.

Distinctive fundamental purification steps from plant sample will be discussed in this chapter including isolation of chloroplasts and mitochondria from leaves as well as from plant tissues rich in phenolic compounds. Meanwhile, we cannot ignore purification of protein from microorganisms. Diverse types of affinity chromatography will be also talked about in this chapter. Perusing can discover informative topics, which will urge them to experience protein purification methodology.

This chapter portrays contingent techniques which are previously or now utilized in the purification of proteins from planta. In particular biochemists have devoted effort to finding convenient methods for protein purification in case they are aiming to obtain a precise characterization and dive inside the target protein. Much advancement has been made, utilizing a lot of partiality systems. Especially, proteins as biomicromolecules differ in their chemical and physical properties. The physical properties demand special techniques for studying, for example, mass spectroscopy, gel filtration, ion exchanger chromatography, electrophoresis, and chromatofocusing, while the chemical properties include the composition, covalent and non-covalent bonds, and solubility.

2. Extraction of protein from plant and yeast

It is impossible to obtain a pure plant protein by using a single-step protocol of purification, as plant tissues contain a wide range of proteins and have rigid cellulosic cell wall and phenolic compound which may cause protein degradation.

The extraction process starts with utilizing a known and appropriate weight of fresh or frozen plant sample under liquid N_2. Homogenize this lot in a prechilled mortar with suitable volume of extraction buffer (3 x of the sample). The extraction buffer type and its pH are distinctive to the sample nature. Some additives should be added to the extraction buffer to improve the obtained extract quality. These additives include:

a. Phenylmethylsulfonyl fluoride (PMSF, dissolved in very small amount of propanol just before extraction) to inhibit protease

b. Flavin compounds (FAD)

c. Dithiothreitol (DTT), to preserve the sulfhydryl group

d. EDTA, as chemical chelator especially with phosphate buffer

e. Sodium fluoride to inhibit phosphatase

f. Polyphenylpyrovate (PVP) 25 gm/k gm fresh weight of sample, which is an insoluble compound and binds to phenolic compound present in the sample, the obtained compound discarded by centrifugation

g. 30% glycerol, which may help to stabilize some highly labile proteins

h. Antibiotic such as Hibitane, usually used in protein extracted from rhizomes or underground plant parts in general

Filter the homogenate with nylon mush, then centrifuge at 10,000 g for 10 min using cooling centrifuge. Repeat the process of extraction and centrifugation three times to ensure complete extraction of all proteins in the tissue. Nearly a similar technique was followed in extraction of Rubisco from wheat leaves [1], protein from *Peganum harmala* seeds [2], and chickpea seeds [3]. Total protein content ought to be determined in the crude extracts [4], as well as the activity if the target protein is an enzyme.

In the case of a microorganism, e.g., yeast, a Yeast Protein Extraction Buffer Kit may be used [5]. This buffer is based on organic buffering agents, which utilize mild nonionic detergents, and a proprietary combination of various salts and agents to enhance extraction and stability of proteins. A ready-to-use Zymolyase preparation is also provided.

3. Concentration of crude extract

3.1 Ammonium sulfate precipitation

To concentrate or reduce the total volume of crude extract, solid ammonium sulfate was added to the crude extracts to bring the final concentration to 70% (w/v) or follow up **Table 1** to choose alternative concentration. After complete

	Final percent saturation to be obtained																
	20	25	30	35	40	45	50	55	60	65	70	75	80	85	90	95	100
Starting percent saturation	Amount of ammonium sulfate to add (gram) per liter of solution at 20°C																
0	113	144	176	208	242	277	314	351	390	430	472	516	561	608	657	708	761
5	85	115	146	179	212	246	282	319	358	397	439	481	526	572	621	671	723
10	57	86	117	149	182	216	251	287	325	364	405	447	491	537	584	634	685
15	28	58	88	119	151	185	219	255	293	331	371	413	456	501	548	596	647
20	0	29	59	89	121	154	188	223	260	298	337	378	421	465	511	559	609
25		0	29	60	91	123	157	191	228	265	304	344	386	429	475	522	571
30			0	30	61	92	125	160	195	232	270	309	351	393	438	485	533
35				0	30	62	94	128	163	199	236	275	316	358	402	447	495
40					0	31	63	96	130	166	202	241	281	322	365	410	457
45						0	31	64	98	132	169	206	245	286	329	373	419
50							0	32	65	99	135	172	210	250	292	335	381
55								0	33	66	101	138	175	215	256	298	343
60									0	33	67	103	140	179	219	261	305
65										0	34	69	105	143	183	224	267
70											0	34	70	107	146	186	228
75												0	35	72	110	149	190
80													0	36	73	112	152
85														0	37	75	114
90															0	37	76
95																0	38

Table 1.
Quantities of ammonium sulfate required to reach given degrees of saturation at 20°C [1].

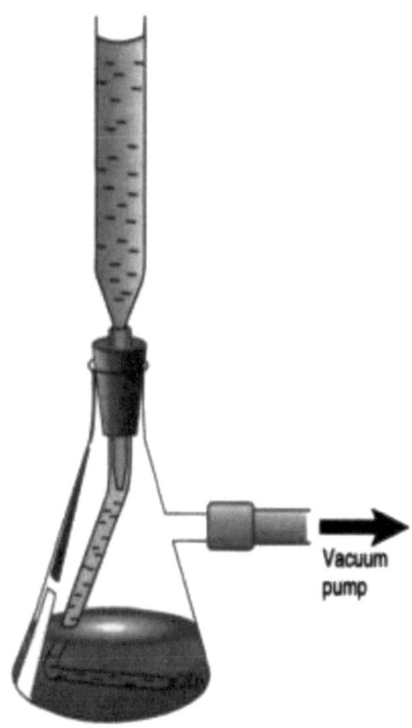

Figure 1.
Experimental arrangement for concentration by forced dialysis [1].

dissolving the added solid ammonium sulfate, the mixture was allowed to stand at 4°C for 24 h, and the precipitate was collected by centrifugation at 10,000 g for 10 min using cooling centrifuge. The precipitated proteins were dissolved in the least amount of extraction buffer. To get rid of excess ammonium ions, dialysis was carried out against the same buffer (see **Figure 1**).

3.2 Fractional precipitation with acetone

The optimum precipitation range for the particular protein fraction is between 37.5 and 50% (v/v). Appropriate and known volume of cruse protein extract is chilled in an ice-salt bath. For each 1 ml of protein solution, add 0.60 ml of acetone (dropwise with constant stirring). After addition of acetone is completed, continue stirring for 10 min with constant control of temperature. The developed precipitate is collected by centrifuging the acetone-protein mixture at 3000 *x*g for 10 min. The precipitated protein is removed and recovered in least volume of extraction buffer. The supernatant volume is measured and 0.25 ml of acetone/ml of protein solution is added further. As in the previous, the precipitated protein by centrifugation is dissolved. To drain off the remaining acetone, the centrifuge tubes over filter paper are inverted. The pellet is suspended in small volume of buffer, and it is kept for the next step of purification [6].

4. Ion-exchange chromatography

4.1 Principles of chromatography

When a sample is applied to any two-phase system (e.g., liquid-liquid, liquid-solid), a molecule may partition between the two phases. With partition coefficient [7],

$$K = C_s/Cm$$

where C is sample concentration in separation phase (s) and in mobile phase (m). When a mixture of several components is applied to such two-phase system, each component of the mixture will have its own individual partition coefficient.

4.2 Components of liquid chromatography

The chromatography system has two components: one is buffer (represents the mobile phase), and the other is hydrated polymers. Buffers should be degassed and filtered before using, especially in the case of high-resolution chromatography. It pumped from a reservoir onto the column. In FPLC and HPLC, high-precision pumps drive mobile phase through the system. The solid phase should be mechanically stable, chemically inert, and widely ranged in their type, efficiency, and

Packing	Composition	Application
DEAE/CM-cellulose	Polysaccharide (cellulose)	Ion exchange chromatography, especially early in purification scheme
Glutathione-agarose	Polysaccharide (agarose)	Affinity chromatography and purification of GST fusion proteins
IDA-agarose	Polysaccharide (agarose)	IMAC (see Section 2.4.6)
Sephacryl S-300	Polysaccharide (dextran/ bis acrylamide)	Gel filtration of proteins in Mass range 10–1500 kDa
Sephadex G-25	Polysaccharide (dextran)	Desalting of protein extracts by gel filtration
Superose 12	Polysaccharide (agarose)	Gel filtration FPLC in the Mass range 1–300 kDa
C-18 Silica	Silica	Reversed-phase HPLC of tryptic peptides
POROS	Poly(styrene-divinylbenzene)	Perfusion chromatography
DEAE/CM-MemSep	Polysaccharide (cellulose)	Membrane-based ion exchange chromatography of proteins

Table 2.
Some stationary phases used in chromatography [7].

application. **Table 2** shows some of the widely used stationary phases; all of them are commercially attainable packed columns.

By the beginning of the chromatographic separation method, the loaded sample is eluted from the column using the mobile phase. This may be achieved in three ways:

a. Continuous flow elution (mobile phase composition and flow rate remain constant).

b. Batch flow elution (the adsorbed sample is selectively eluted by using a range of mobile phase) and stepwise introduction of different mobile phases varied in pH, polarity, and salt concentration.

c. Gradient elution (the adsorbed material separated by two or more mobile phase compositions) is continuously varied.

The chosen way for elution is depending on the behavior of sample components on the stationary phase (**Figure 2**).

4.3 Types of ion chromatography and elution ways

Amino acid sequence determines the net charge of protein. Ion-exchange chromatography separates proteins either on the basis of their charge type (cationic or anionic) or charge strength (e.g., strongly anionic from weakly anionic). Like most column chromatography techniques, ion-exchange chromatography requires a stationary phase which is usually composed of insoluble, hydrated polymers, such as cellulose, dextran, and Sephadex. **Table 2** indicates some of the commonly used ion exchangers [1]. The main concept is that the anion like a negatively charged protein exchanges with chloride ions, while the cation like a positively charged protein exchanges with potassium or sodium ions. This process can later be reversed by washing with chloride ions in the form of NaCl or KCl solution. After loading the crude solution that contained the target protein, some of the proteins are bounded; however, others are not. Elution is carried out by washing with buffer which contained an

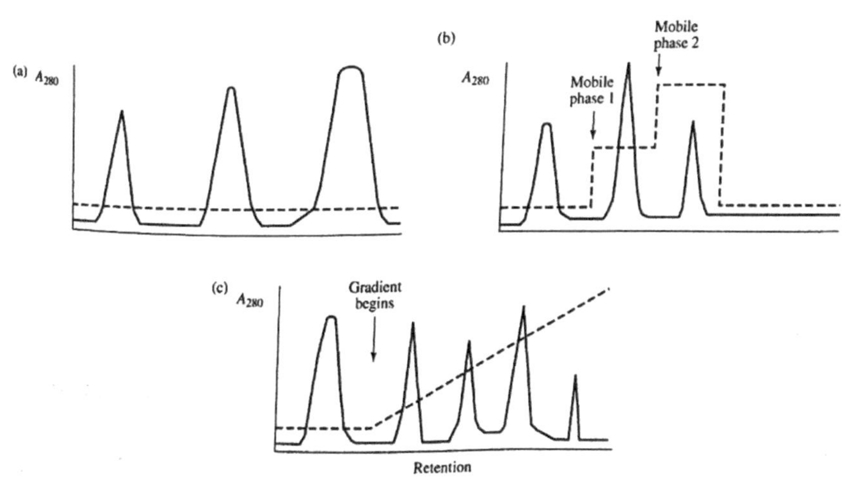

Figure 2.
Elution from stationary phase. Mobile phase is shown as dashed lines. (a) Continuous flow elution. (b) Batch flow elution. (c) Gradient elution, where the adsorbed material is separated by the application of two or more buffers in which mobile phase composition is continuously varied [7].

electrolyte. During elution step, weakly bound proteins are firstly removed, followed by more strongly bound proteins. The opposite is correct for the cation, which like a positive charged protein exchanges with sodium or potassium ion. As an example, in the purification of uricase from soybean, DEAE-cellulose which is an anion exchanger is used. DEAE-cellulose column was prepared and pretreated as recommended by Peterson and Sober [8]. Column was equilibrated with 50 mM phosphate buffer pH 7.5, 1.0 mM EDTA, and 10.0 M mercaptoethanol. The protein was eluted with a stepwise gradient from 0.0 to 0.5 M KCL solutions prepared in the same buffer used in column equilibration. This column was successfully used for purification of uricase extracted from leaves and root nodules of soybean [9]. It is worth to measure uricase activity and protein content in each eluted fraction. Soluble protein was routinely determined spectrophotometrically at 280 nm. Separated peaks may be collected by fraction collector. For more perfection, some researcher may also use 205–220 nm to ensure not missing proteins or peptides which have poor aromatic residues. Elution profile is represented by constructing a plot between detector signal and the retention of sample contents (eluted fractions vs. OD_{280}).

5. Gel filtration chromatography

5.1 For protein purification

Shape and mass of proteins are unique. Gel filtration is also known as size exclusion in which the stationary phase is made up of a gel consisting of beads, which have narrow and fixed size pores (**Figure 3**). So, the proteins below the size pore will diffuse out from the column. Each type of gel filtration has certain cutoff value. By loading the sample, the smaller components pass through the bed and retain faster than the larger ones. The greatest component in the size is eluted freely from the column just after the void volume. Hence, the fraction range will depend uniformly on the population of beds in the column.

Sephacryl S-100 and Sephadex G-100 columns were used extensively for this step. These columns were equilibrated with extraction buffer such as 50 mM phosphate buffer pH 7.5, 1.0 mM EDTA, and 10.0 mM mercaptoethanol. The protein was eluted with the same buffer. Enzyme activity and/or protein content should be assayed for each of the eluted fractions.

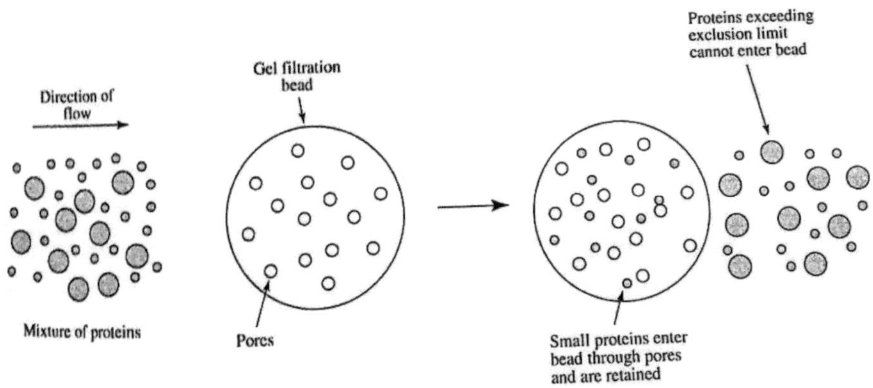

Figure 3.
Gel filtration chromatography of proteins. The proteins smaller than the bead exclusion limit may enter the bead. Larger proteins are excluded. Therefore, proteins elute in inverse order of native mass [7].

5.2 For mass determination

The relative native molecular mass of purified proteins was determined after fractionation on Sephadex G-100 (23 x 1.5 cm) equilibrated with 50 mM phosphate buffer pH 7.5, 1.0 mM EDTA. The marker proteins of known low molecular mass used such as:

aprotinin (6500), cytochrome c (12,400), carbonic anhydrase (30,000), and albumin bovine serum (66,000) Da. The void volume (V_o) was determined with dextran blue (2,000,000) Da. Purified enzyme samples of 2.0 ml volume contained about 2 mg protein/ml was applied to the same column. The protein was eluted from the column with the same buffer used for equilibration. At a flow rate of 1 ml/2.5 min, 2.5 ml fractions were collected. A calibration curve was constructed by plotting log molecular masses versus V/V_o, where (V) was the elution volume and (V_o) was the void volume.

5.3 For removal of low molecular components

Gel filtration may be used in routine way to remove salts. Sephadex G-10 and Sephadex G-25 are commonly used to desalt samples, and hence, it makes the exchange process more rapid and minimizes the dilution of protein sample.

6. Affinity chromatography

The principle of this technique is to fix a small molecule which is called "ligand" on the column stationary phase just before sample loading (**Figure 4**). Affinity is bispecific recognition and distinguishing of biomolecules. **Figure 5** shows a schematic diagram of the steps involved in an affinity-based separation [10].

There are many types of this technique, the most used of them are:

6.1 Substrate or inhibitor affinity

In this technique an enzyme will be recognized by its substrate or inhibitors or cofactors. As a result of the highly specific nature of this ligand to the target protein, only this protein will be bound to the stationary phase and hence purify with high quality. On the other hand, the other protein molecules will go through the chromatography system. At the end, the target protein will be eluted from the ligand. In the case of closely related proteins or isozymes, a variation of this approach should be applied to create an affinity gradient, where the concentration of ligand is gradually increasing in the mobile phase. Hence, each isozyme is unique in its K_m's for the substrates or K_i's for the inhibitors.

6.2 Metal affinity

In this technique, metals are binding to a stationary phase, and metal-binding regions of proteins bind to this ligand. The most common metal used in this technique is Cu, Zn, Ca, Ni, and Fe which are covalently bound to iminodiacetic acid (IDA) and then to the stationary phase, i.e., agarose or silica. A critical problem faced by the researchers during using this technique is that some amino acids, i.e., histidine, cysteine, and tryptophan, may be gradually leached from the stationary phase due to the weakness of interaction with IDA. Elution of the target protein is usually displayed by decreasing pH gradient.

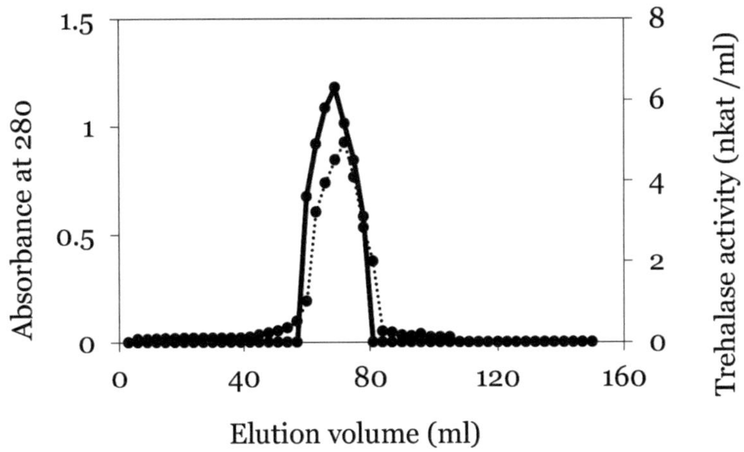

Figure 4.
A typical elution profile for the behavior of trehalase of seeds of T. aestivum *on Sephadex G-200 column (30 × 1.5 cm) equilibrated with 100 mM sodium citrate buffer (pH 5.5). The flow rate was 1.5 ml min*$^{-1}$ *and 3 ml fractions were collected. Solid line indicates the absorbance; the dotted indicated the enzyme activity [9].*

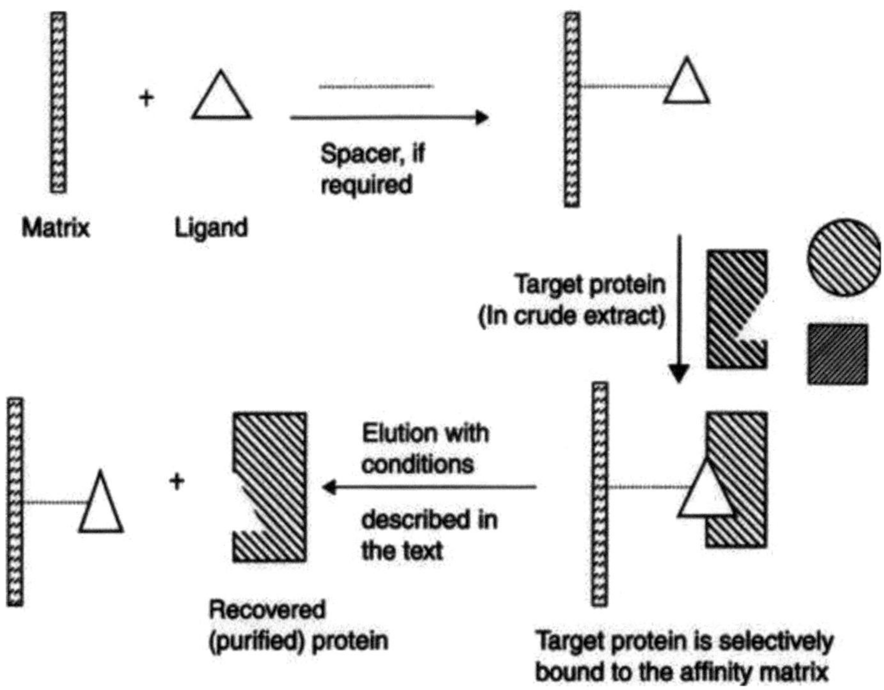

Figure 5.
Schematic diagram of the steps involved in an affinity-based separation [1].

6.3 Immunoaffinity

Immunoaffinity chromatography (IAC) is a type of liquid chromatography that uses a highly specific binding of antibodies for an antigen which is representing the target protein. The application of IAC includes immune-extraction, immune-depletion, chromatographic immunoassays, and post-column immune-detection [11]. These different applications are coming from the tight and selective interactions between antigens and antibodies. Versatile applications of IAC by using either immobilized antigens or antibodies have been used on purification of wild-type and recombinant mutant amidases from *Pseudomonas aeruginosa* [12]. Antibodies are isolated as an antibody class for immunodiagnostics and biopharmaceutics, and the higher levels of purification are beneficial for the diagnostic methods and for the therapeutic applications. With respect to the importance of purification of antibodies, more and more attention has been paid to IAC, which is an efficient protein separation method based on the interaction between target proteins and specific immobilized antibodies [13, 14]. Monoclonal antibodies (MAbs) are useful for the treatment of a wide array of indications including autoimmune diseases, infectious diseases, cardiovascular diseases, transplant rejection, and cancer. In this respect, the IgG class which plays a most pronounced role in the clinical application was extremely purified using specific antibodies (hyperimmune IgG) [15, 16]. With respect to the importance of purification of antibodies, more and more attention has been paid to IAC, which is an efficient protein separation method based on the interaction between target proteins and specific immobilized antibodies. The less cost and low risk as well as the more safety prompted research activities to focus on novel synthetic ligands (synthetic mimic ligands of proteins A and L). Immunodiagnostically useful *Mycobacterium tuberculosis* H37Ra protein antigens ES-31, ES-43, and EST-6 were isolated from detergent soluble sonicate (DSS) antigen using monospecific antibodies by affinity chromatography. ES-31, ES-43, and EST-6 antigens purified from both culture filtrate and DSS antigen showed similar seroreactivity with overall sensitivity 85, 80, and 75%, respectively [17]. Shuai Sheng and Fansheng Kong reviewed the advantages, applications, as well as drawbacks of IAC in the separation and purification of antibodies and antigens [18].

6.4 Fusion tag protein purification

When proteins are expressed recombinantly, additional amino acids, a functional domain, or a whole protein is often appended to aid in the purification. A standout among the most widely recognized combination labels is His, which is a short chain of six to nine histidine buildups (known as the 6xHis or polyHis tag). Application of 6xHis in purification gives 1000-fold of binding specificity than that of IDA-agarose [7]. The later will bind to metal ions such as nickel or cobalt. Another fusion tag, glutathione S-transferase (GST), which binds tightly to reduced glutathione (GSH), is mostly used. Unlike His and GST tags, most of these other tags are called epitope tags because they require specific antibodies (e.g., immobilized anti-HA antibody) for purification. Epitope labels are irregularly used as it is generally far reaching contrasted with straightforward ligand media, for example, nickel or glutathione agarose. In this way, antibody combination labels are broadly accessible for utilization of single-step liking cleaning.

6.5 Heme-tagged protein

The addition of affinity tags to the amino- (N) or carboxyl- (C) terminus of a target protein for purification purposes can dramatically reduce the amount of preparation time, reduce the number of purification steps needed, and increase the yield of pure protein [19–22]. Recently, the combination of heme tag-HIS has been found more convenient as the target protein was visible and easy to be detected. For this purpose, as the heme chromophore grants the fusion protein with an intense red or red brown color, visible detection significantly reduces the time and effort associated with protein expression and handling during all purification and chromatographic steps. Heme tags are composed of a heme-binding peptide (HBP) that is linked to the heme chromophore by two stable thioether bonds [23]. HBPs are used for HIS affinity purification. HBPs are typically small, consisting of 14–20 amino acid residues. The critical element of a HBP is a Cys-X-X-Cys-His (CXXCH) heme-binding motif within the sequence, with X representing any amino acid residue [24]. The basis of the heme-tag-HIS affinity purification method is the reversible coordination between the heme-iron open coordination site and the side chain of histidine immobilized on Sepharose beads. Later on, the target protein can be eluted from the column using an imidazole-containing buffer, a low pH (≤5) buffer, or a high pH (≥8) buffer.

6.6 Avidin-biotin systems

Biotin or vitamin H is a small molecule (MW 244.3) that is present in humble amounts in all living cells. The acidic side of this vitamin can be incorporated with various reactive groups. Once biotin is attached to a molecule, the molecule can be captured for detection, immobilization, or affinity purification using conjugates or supports based on avidin or streptavidin proteins, which bind strongly and explicitly to the biotin group. Native and recombinant derivatives of avidin and streptavidin proteins are promptly accessible in a wide assortment of altered, labeled, and immobilized forms. The "avidin-biotin system" has been adapted for use in many kinds of research applications for detection or purification. As the avidin-biotin affinity interaction is so strong, therefore it is usually impractical to elute biotinylated targets that have been captured to immobilize avidin or streptavidin support. However, modified versions of biotin labeling reagents have been developed, such as cleavable biotin, iminobiotin, and desthiobiotin; these provide readily reversible interactions with streptavidin, making them useful tools for soft-release applications.

6.7 Cellulose binding model

Cellulose binding module (CBM) has been distinguished as a potential tool for the fast and easy assembling of recombinant proteins [21, 25]. CBM is an attractive affinity tag for protein purification because of its high capacity and specific adsorption to cellulose. In addition, it can be efficiently adsorbed to cellulose in most buffers and eluted under non-denaturing conditions. There are many advantages of this technique such as low cost, appropriate physical properties, applicable for most proteins, and high stability in the presence of buffers. All of these make it an ideal matrix for large-scale affinity. Moreover, cellulose is easily procured and safe for many pharmaceutical and human purposes. The method of purification of the recombinant protein with a CBM3 tag was developed in both *Escherichia coli* and yeast and has been proven successful in purifying several proteins [26, 27]. This study introduced the recombinant protein expressed in yeast. EGFP was used to demonstrate the well-ordered purification procedure. A self-cleaving intein-based tag removal method for CBM3 was also introduced [28].

6.8 Affinity precipitation

A novel affinity precipitation developed by Arnold and Chen [6] was very promising. One-step binding and precipitation resulted in >95% recovery yield directly from crude extract and a 22.7-fold purification, giving a specific activity of 420 U/mg. The soybean peroxidase isolated using this affinity precipitation meets or exceeds the quality specifications of reagent grade products. The molar ratio of ligand to target fucosylated protein is key for the recovery yield.

Affinity precipitation is an alternative approach with the potential to overcome the challenges associated with affinity chromatography, as it retains the specific interaction of affinity ligand with the protein of interest while avoiding column operation [29].

7. Conclusion

As there is no restricted catalog for purification steps of proteins to follow it up, the researcher should design its own protocol according to the aimed target and the final quality that is being looked for. Gel filtration and ion exchangers are suitable for broad mass and low-cost aim. However, affinity chromatography with its versatile types is the only technique for highly purified and selective protein yield. All purification techniques have the same major steps that include equilibration of a column, adsorption of a sample, washing to remove unbound materials, and elution of bound protein target protein and ended with regeneration of the media.

Author details

Hanaa E.A. Amer[1,2]

1 Department of Botany and Microbiology, Faculty of Science, University of Cairo, Giza, Egypt

2 Department of Biological Science, Faculty of Science, University of Jeddah, Jeddah, KSA

*Address all correspondence to: hanaaelbadawwy@gmail.com

References

[1] Doonan S, editor. Protocols PP. In: Protein Purification Protocols. Humana Press Inc.; 1996

[2] Ahmed HE, ElZahab HA, Alswiai G. Purification of antioxidant protein isolated from *Peganum harmala* and its protective effect against CCl_4 toxicity in rats. Turkish Journal of Biology. 2013;**37**(1):39-48

[3] Kord M, Youssef E, Ahmed HE, Qaid E. Purification and characterization of trehalase from seeds of chickpea (*Cicer arietinum L.*). Turkish Journal of Biology. 2013;**37**(6):661-669

[4] Lowry OH, Rosenbrugh NJ, Farr AL, Randall RJ. Protein measurement with the Folin phenol reagent. *The Journal of Biological Chemistry*. 1951;**193**:265-275

[5] Handbooks from GE Healthcare Life Sciences For more information refer to www.gelifesciences.com/handbooks

[6] Arnold L, Chen R. One-step non-chromatography purification of a low abundant fucosylated protein from complex plant crude extract. Bioengineering. 2015;**2**:249-263

[7] Sheehan D, editor. Physical Biochemistry: Principles and Applications. 2nd ed. Wiley; 2010. pp. 11-51

[8] Peterson EA, Sober HA. Column chromatography of proteins: Substituted cellulose. In: Colowick SP, Kaplan NO, editors. Methods in Enzymology. Vol. 5. New York: Academic Press; 1962. pp. 3-27

[9] Ahmed HE. Purification and characterization of uricase from nodulated soybean plants treated with IAA [PhD thesis]. Cairo University; 2005

[10] Gupta MN, Guptaand Roy I. Affinity-based separation: An overview. In: Gupta MN, editor. Methods for Afinity-Based Separations of Enzymes and Proteins. Basel: Springer; 2002. pp. 1-15

[11] Jackson AJ, Sobansky MR, Hage DS. Principles and applications of immunoaffinity chromatography. In: Antibodies Applications and New Development. 2012. pp. 156-174

[12] Martins S, Lourenço S, Karmali A, Serralheiro ML. Monoclonal antibodies recognize conformational epitopes on wild-type and recombinant mutant amidases from *Pseudomonas aeruginosa*. Molecular Biotechnology. 2007;**37**:136-145

[13] Grønborg M, Kristiansen TZ, Stensballe A, Andersen JS, Ohara O, Mann M, et al. A mass spectrometrybased proteomic approach for identification of serine/threoninephosphorylated proteins by enrichment with phospho-specific antibodies: Identification of a novel protein, Frigg, as a protein kinase A substrate. Molecular & Cellular Proteomics. 2002;**1**:517-527

[14] Steen H, Kuster B, Fernandez M, Pandey A, Mann M. Tyrosine phosphorylation mapping of the epidermal growth factor receptor signaling pathway. The Journal of Biological Chemistry. 2002;**277**:1031-1039

[15] D'Agostino B, Bellofiore P, De Martino T, Punzo C, Rivieccio V, Verdoliva A. Affinity purification of IgG monoclonal antibodies using the D-PAM synthetic ligand: Chromatographic comparison with protein A and thermodynamic investigation of the D-PAM/IgG interaction. Journal of Immunological Methods. 2008;**333**:126-138

[16] Verdoliva A, Pannone F, Rossi M, Catello S, Manfredi V. Affinity

purification of polyclonal antibodies using a new all-D synthetic peptide ligand: comparison with protein A and protein G. Journal of Immunological Methods. 2002;**271**:77-88

[17] Upadhye V, Saha-Roy S, Shende N, Kumar S, Harinath BC. Isolation of mycobacterium tuberculosis protein antigens ES-3 1, ES-43 and EST-6 of diagnostic interest from tubercle bacilli by affinity chromatography. Indian Journal of Experimental Biology. 2007;**45**:599-602

[18] Sheng S, Kong F. Separation of antigens and antibodies by immune-affinity chromatography. Pharmaceutical Biology. 2012;**50**:1038-1044

[19] Young CL, Britton ZT, Robinson AS. Recombinant protein expression and purification: A comprehensive review of affinity tags and microbial applications. Biotechnology Journal. 2012;7:620-624

[20] Chelur D, Unal O, Scholtyssek M, Strickler J. Fusion tags for protein expression and purification. BioPharm International. 2008;**21**(Suppl. 6):38-46

[21] Arnau J, Lauritzen C, Petersen GE, et al. Current strategies for the use of affinity tags and tag removal for the purification of recombinant proteins. Protein Expression and Purification. 2006;**48**:1-13

[22] Lichty JJ, Malecki JL, Agnew HD, et al. Comparison of affinity tags for protein purification. Protein Expression and Purification. 2005;**41**:98-105

[23] Zhong M, Molday LL, Molday RS. Role of the C terminus of the photoreceptor ABCA4 transporter in protein folding, function, and retinal degenerative diseases. The Journal of Biological Chemistry. 2009;**284**(6):3640-3649. DOI: 10.1074/jbc.M806580200

[24] Loewen CJ, Moritz OL, Molday RS. Molecular characterization of peripherin-2 and rom-1 mutants responsible for digenic retinitis pigmentosa. The Journal of Biological Chemistry. 2001;**276**(25):22388-22396. DOI: 10.1074/jbc.M011710200 M011710200

[25] Przybycien TM, Pujar NS, Steele LM. Alternative bioseparation operations: Life beyond packed-bed chromatography. Current Opinion in Biotechnology. 2004;**15**:469-478

[26] Banki MR, Gerngross TU, Wood DW. Novel and economical purification of recombinant proteins: Intein-mediated protein purification using in vivo polyhydroxybutyrate (PHB) matrix association. Protein Science. 2005;**14**(6):1387-1395. DOI: 10.1110/ps.041296305

[27] Gallagher SR. One-dimensional SDS gel electrophoresis of proteins. Current Protocols in Molecular Biology. 2012. DOI: 10.1002/0471142727.mb1002as97. Chapter 10:Unit 10 12A

[28] Wang D, Hong J. Purification of a recombinant protein with cellulose- binding module 3 as the affinity tag. In: Giannone RJ, Dykstra AB, editors. Protein Affinity Tags. Humana Press; 2014. pp. 35-46. DOI: 10.1007/978-1-4939-1034-2

[29] Linné-Larsson E, Galaev I, Lindahl L, et al. Affinity precipitation of Concanavalin A with p-aminophenyl-α-D-glucopyranoside modified Eudragit S-100. Bioseparation. 1996;**6**:273-282

Section 2

Proteomics Applications

Chapter 4

Proteomic Analysis of β-Thalassemia/HbE: A Perspective from Hematopoietic Stem Cells (HSCs)

Saranyoo Ponnikorn, Siripath Peter Kong, Sasipim Thitivirachawat, Chanawin Tanjasiri, Sumalee Tungpradabkul and Suradej Hongeng

Abstract

β-Thalassemia/HbE is highly prevalent in Southeast Asian countries, especially Thailand. It is a severe hereditary anemia disease involving ineffective erythropoiesis in the bone marrow and peripheral tissues. The excess of alpha globin and iron overload contribute to elevated oxidative damages leading to a premature cell death of erythroid cells and a diminished terminal differentiation of reticulocytes. Although proteomic approach would gain a comprehensive picture of the complex pathophysiology of human bone marrow and hematopoietic stem cells (HSCs), obtaining sufficient clinical specimens remains an important issue. The employment of mass spectrometry (MS)-based proteomic profiling could overcome these constrains and provide useful insights into the cellular constituents and microenvironment in bone marrow milieu. In this chapter, we summarize the comparative proteomic studies analyzing $CD34^+$/HSCs and bone marrow niche proteins. Under ineffective erythropoiesis, in-depth analyses of various proteome profiles revealed many of which have putative functions. Importantly, dysregulated cell death and survival signaling pathways could explain the deleterious pathogenesis of β-thalassemia/HbE.

Keywords: β-thalassemia/HbE, hematopoietic stem cells (HSCs), ineffective erythropoiesis and proteomics

1. Introduction

Erythrocytes contain the hemoglobin (Hb) protein that carries oxygen from the lungs to the rest of the body. During the development, the hemoglobin structure gradually changes. The fetal hemoglobin HbF ($\alpha_2\gamma_2$) is a major hemoglobin in the human fetus during the last 7 months of development in the uterus. The adult hemoglobin contains two components, the major hemoglobin, Hb A ($\alpha_2\beta_2$), and the minor hemoglobin, Hb A2 ($\alpha_2\delta_2$) [1, 2]. The composition of heme and heterodimeric forms of globin protein, α-globin and β-globin, is encoded by the α gene and the β gene, which are in chromosome 16 and 11, respectively. The inherited hemoglobin disorder is a common monogenic disease that could be identified into two main groups, the hemoglobinopathy which is abnormal structure hemoglobin variants and the recessive inheritance allele which is the defective hemoglobin production, the thalassemia syndromes [3, 4]. More than 200 point mutations and 80 different deletions within the β-globin-encoding gene are related to inherited hemoglobin disorder of β-thalassemia that represent in population worldwide [5]. In Southeast Asia, especially Thailand, hemoglobin E (HbE) thalassemia accounts for about one-half of all cases, with the highest frequencies observed. Approximately 35,000 patients are living with β-thalassemia syndrome in Thailand, 7% with a β-thalassemia trait, and 17% with an HbE trait within a population of 65 millions of Thai people [6].

Pathophysiology of β-thalassemia/HbE is strongly associated with the excess of unmatched α-globin chain and mutation of hemoglobin variants which are apparently observed inside the bone marrow, as well as peripheral tissues including the liver and spleen [7, 8]. These phenomena are believed to be the etiological factors of ineffective erythropoiesis. However, these conditions are found in certain hereditary blood disorders and are often associated with β-thalassemia/HbE [9]. This disease is characterized by apoptosis of erythroid precursors and inhibited erythroid differentiation and maturation [10]. These lead to an insufficient number of red blood cells. Consequently, a decreased oxygen-carrying capacity of the blood is tightly linked to the compensatory mechanism of the elevated erythropoietin (EPO) production that later causes the deleterious effects [11].

Erythropoiesis is a complex orchestrated process involving self-renewal, proliferation, differentiation, and mobilization of HSCs that gives rise to erythroid lineages [12]. These processes are governed by heterogeneous regulatory factors, cytokines, signaling molecules, and various cellular interactions. Erythropoiesis is extensively studied in mice model, although there are several species-specific differences of erythroid differentiation [13]. Many in vitro studies focused on cultures of human hematopoietic stem cells (HSCs), and erythroid progenitor cells have been elucidated the pathogenesis of β-thalassemia/HbE [14–20]. However, studying of homeostatic imbalance inside the bone marrow regarding ineffective erythropoiesis is challenged by limited human samples.

The emergence of "OMIC" approach, especially proteomics, has accelerated our understanding of pathobiology of the β-thalassemia/HbE disease, which will have a substantial impact in medicine. Technological advances of proteomic analysis tools, for example, mass spectrometry, allow high sensitivity, reduced sample requirement, increased throughput, and dissect posttranslational modifications of limited clinical specimens [21, 22]. The use of these technologies continues to expand substantially, particularly to meet the need for better diagnostics and to shorten timeline of effective therapy development. Interestingly, in combination with optimized ex vivo HSCs to erythroid culture or freshly isolated samples, MS-based proteomic approach has revealed a more comprehensive understanding of ineffective erythropoiesis in β-thalassemia/HbE.

2. Pathophysiology of β-thalassemia/HbE

β-Thalassemia is characterized by insufficient β-globin chain production due to decreased production (β^+) or null production (β^0) of β-globin chain, which is the main component of adult hemoglobin. This condition leads to variable clinical symptoms. β-Thalassemia major is characterized by severe anemia. Individuals with homozygous β-thalassemia may develop β-thalassemia major or β-thalassemia intermedia. The concentration of hemoglobin could distinguish types of β-thalassemia

[3, 4, 23, 24]. β-Thalassemia major has low levels of hemoglobin of 3–4 g/dL, whereas the mild-to-moderate anemia show hemoglobin concentration at 7–10 g/dL. Clinical complications of β-thalassemia major include growth retardation, pallor, jaundice, poor musculature, genu valgum, hepatosplenomegaly, and leg ulcers. A severe anemia is accompanied by ineffective erythropoiesis with bone expansion and extramedullary erythropoiesis in the liver, spleen, and other sites such as paravertebral masses. These lead to the skeletal changes including deformities in the long bones of the legs and typical craniofacial changes [3, 25]. Typical treatment for β-thalassemia major is primarily blood transfusion which causes another complication, posttransfusional iron overload. In some cases, chelation therapy could be implemented.

β-Thalassemia/HbE has extremely diverse clinical phenotypes because of various genetic lesions of the gene affecting different impaired globin chain synthesis and abnormal forms of hemoglobin variants [3, 4]. The genetic mutation of the β-globin-encoding gene at codon 26 (GAG to AAG) contributes to the activation of cryptic splice site that cause abnormal mRNA processing. The β^E-chain are produced at a reduced level resulting in a mild β-thalassemic phenotype [26]. Each year in Thailand, approximately 3000 infants are born with β-thalassemia/HbE either a mild β-thalassemia or a severe transfusion-requiring β-thalassemia major. HbE/β-thalassemia patients have remarkable variation in terms of severity reflecting the heterogeneity of the β-thalassemic mutations in the HbE gene and other modulation factors. Newborns carrying HbE/β-thalassemia mutations are usually asymptomatic because they rely on the HbF hemoglobin rather than the HbE. After 6–12 month of birth, anemia with abdominal enlargement due to hepatosplenomegaly could be observed [27–30]. Although the defect in β-globin production could be explained by the mutations or the absence of the β-globin gene or other modification factors, the variable severity of β-thalassemia is also involved in an imbalance of β- and α-chains, particularly in maturing erythroid cells [31]. The accumulating unmatched free α-globin chain is unstable and could precipitate in the erythroid precursors in the bone marrow as well as in peripheral blood. A severe accumulation of the α-chain forms hemichromes which cause uncontrollable oxidative damages in erythroid precursor and erythroid cells. The hemichromes can generate reactive oxygen species (ROS) that potentially damage important proteins, DNA, and other cellular components and trigger signaling cascades of programmed cell death [10, 32–36]. The destruction of erythroid cells is an onset of dramatic amelioration cycles inside the bone marrow and peripheral organ due to the ineffective erythropoiesis in β-thalassemia/HbE patients.

3. Ineffective erythropoiesis

Erythropoiesis is a progressive orchestrated process involving heterogeneous regulatory factors such as cytokines, signaling molecules, and various cellular interactions which play important roles in self-renewal, proliferation, differentiation, and mobilization of HSCs to become erythroid lineage [13, 37]. The ineffective erythropoiesis is the hallmark pathophysiological condition having seen in β-thalassemia/HbE. In β-thalassemia/HbE, the oxidative damage caused by the accumulation of hemichromes and the generation of ROS in red blood cells (RBCs) and erythroid progenitors are an etiologic factor perturbing the homeostasis of erythropoiesis [8]. In vitro and in vivo studies of mice model and ex vivo studies of human HSCs and erythroid progenitors have elucidated the molecular pathogenesis of the ineffective erythropoiesis of β-thalassemia. The excess of α-chain is believed to accelerate apoptotic cell death in erythroid precursors [10], while the inhibited processes involving differentiation of erythroid maturation have been observed during polychromatophilic stage [34]. Both detrimental processes lead to an insufficient number of RBCs and decreased oxygen-carrying capacity of blood and ultimately cause hypoxia condition. The lack of mature RBCs stimulates a compensatory mechanism of erythropoiesis that triggers the overproduction of EPO from the kidney. The increased EPO could induce erythropoiesis in the bone marrow by inducing proliferation of HSCs and erythroid progenitors. The EPO also triggers survival mechanism under stress condition and supports differentiation of mature RBCs. However, the dramatic oxidative damages inside the bone marrow are the key factor that generates a homeostatic imbalance of erythropoiesis. This is linked to an erythroid expansion leading to extramedullary hematopoiesis [37]. Dysregulated hemoglobin synthesis and hemolysis are major factors of the ineffective erythropoiesis and determine the severity of the disease [38]. Another clinical complication is an iron overload in severe patients that are treated by regular blood transfusions which is not required for thalassemia intermedia patients. Blood transfused patients usually develop elevated body iron load due to an increased gastrointestinal iron absorption. Consequently, the accumulation of iron also causes oxidative stress because they produce highly reactive hydroxyl radicals that could not be naturally degraded by enzymatic reaction. Altogether, the accumulation of unstable free α-globin, heme, and iron all contributes to destruction of cellular compartments in the bone marrow of β-thalassemia patients [7, 39, 40]. Previous studies attempted to identify molecular and cellular mechanisms associated with the ineffective erythropoiesis in β-thalassemia patients. However, the bone marrow is tightly linked to other vital organs and complicates the deciphering of ineffective erythropoiesis of β-thalassemia.

Although there are well-established genetic manipulation techniques in model organisms such as mice and zebrafish and extensive enrichment methods of specific cell stage of hematopoietic progenitors make functional study of hematopoiesis possible, many components of the process act differently in human due to the evolutionary distance of hematopoietic genes, transcription, and epigenetic regulations [13]. Ex vivo human model using primary cell cultures of HSCs obtained from the peripheral blood, cord blood, and bone marrow or fetal liver of donor remains a gold standard of erythropoiesis study. Notably, involving multifactors could be determined in very proliferated and differentiated cells in different culture systems [41]. However, in expansion and differentiation steps, primary cell culture system may not provide sufficient amounts of analytes for some downstream techniques such as Western blot. While most techniques have their limitation to systematically analyze the bone marrow, its microenvironment and other neighboring cellular interactions during pathological conditions of ineffective erythropoiesis of β-thalassemia/HbE. Alternative approaches could be implemented to analyze the limited isolated primary cells of thalassemia patients and potentially other dyserythropoietic diseases. Here, we discuss advantages of OMIC approaches especially proteomics for studying molecular pathogenesis of β-thalassemia/HbE.

4. Clinical proteomics of β-thalassemia/HbE

One of the goals of proteomic study is to characterize the flow of information through complex protein networks in a particular disease model, where the changes of proteins in the whole proteome can be identified in the pathophysiological conditions at different disease developmental stages. Comparative proteome profiling of patient and healthy subjects may help to provide new clinical conceptualization for systemic pathophysiology including root causes, biomarkers, and applicable treatment regimes. Clinical proteomics is an exciting sub-discipline of proteomics that will integrate the application of proteomic technology to identify problems at the bedside [42]. Conventional analysis methods, such as Western blot, are difficult because of limited amounts of clinical samples in many disease models. In the field of biomedical research, nonetheless, the cause of the disease could not be thoroughly explained due to the complexity of various molecular and cellular pathways in many types of cells and tissues. Therefore, elucidation of protein network alteration in clinical samples is an advantage strategy for deciphering interaction between proteins and their expression profiles which are related to cellular physiology [43]. Despite high-throughput genomic technology can identify differential gene expression in disease samples, proteome alterations still occur in many ways not predictable from genomic analysis. A better understanding of these alterations will have a substantial impact in medicine [44, 45]. Moreover, clinical proteomic is a very promising tool for identifying various posttranslational modifications, such as phosphorylation and glycosylation. The recent advance proteomic analysis provides high sensitivity, reduced sample requirement, increased throughput, and is able to uncover various posttranslational modifications which are available for disease-related applications [46]. The use of these technologies will likely expand substantially, particularly to meet the need for better diagnostics and to shorten the path for developing effective therapy.

Proteomic analysis of blood samples has become an interesting method in hematological studies. The extremely rich spectrum of proteins in blood and blood components serves various cellular activities including coagulation, transport, immune system and cell signaling, as well as by-products of cellular damage and proteins from other tissues [47]. Whereas, the identification of originality related hematopoiesis from HSCs in some clinical respects such as leukemia and anemia are very challenging [48]. Since the key candidate proteins belonging to any disease status are in the low-abundance level, to overcome these problems, protein samples are fractionated to allow effective detection and quantification by mass spectrometry (MS) [21]. Different protein separation methods influence MS results. Many previous clinical proteomic studies of β-thalassemia/HbE employed gel-based technique and shotgun proteomic to analyze posttranslational modifications of the thalassemic proteome. Most comparative protein profiles of β-thalassemia/HbE between patients with normal subjects from various types of clinical specimens including plasma serum, platelets, platelet-free plasma-derived microvesicles, and circulating extracellular vesicles (EVs) have been reported [49, 50]. Interestingly, a few studies have directly investigated proteomic changes inside the bone marrow or isolated the HSCs of thalassemic samples, where are the major site of erythropoiesis and could identify relevant proteins in the ineffective erythropoiesis. HSCs were isolated from bone marrow cells and peripheral blood using antibody specific to CD34, the transmembrane glycoprotein. Both peripheral blood- and bone marrow-derived $CD34^+$/HSCs showed differentially regulated proteomes during ineffective erythropoiesis in β-thalassemia/HbE [15, 18]. Recently reports have revealed the nature of the bone marrow plasma proteome and the roles of microenvironment in the bone marrow niche and linked to the dyserythropoiesis of β-thalassemia/HbE [51].

4.1 Gel-based technique

Gel-based proteome analysis method involves a separation of proteins using either one-dimensional (1-D) or two-dimensional electrophoresis (2-DE). Since 2-DE separates proteins by their isoelectric points and by molecular weight, it is a standard choice for protein separation. After the proteins are resolved, the gel is visualized by staining with visible dyes, such as Coomassie Blue or silver nitrate or with fluorescent dyes such as Sypro Ruby. The gel is imaged and analyzed by software to identify protein spot that show altered intensity. Selected protein spots are excised and subjected to in-gel digestion typically by trypsin, resulting in peptides that can be further analyzed by mass spectrometry (MS) or tandem mass spectrometry (MS/MS). One of the drawbacks of the conventional 2-DE is gel-to-gel variability. For many proteomic studies, multiple gels of many samples have to be run, and spots at a certain location have to match between gels to reduce variability, but this still cannot fully eliminate the problem. Apart from variability, 2-DE also has poor reproducibility and provides relatively lower sensitivity. To circumvent these problems, two-dimensional difference gel electrophoresis (2-DIGE) has been developed. By labeling different samples with different fluorescent dyes and separating them in the same gel, 2-DIGE enables better comparison of up to three protein samples [22, 52, 53]. However, there are some proteins cannot be resolved on 2-D gels due to their physicochemical properties. These include highly basic or acidic proteins, proteins with extremely high or low molecular weight, and membrane proteins that are inherently insoluble in gel matrices [21].

Two-dimensional gel electrophoresis (2-DE) coupled with mass spectrometry or tandem mass spectrometry (MS and MS/MS) has been used for the investigation of β-thalassemic proteomes. The majority of differential expressed proteins between β-thalassemia patients and normal clinical specimens have been identified and characterized their functions in different metabolic processes in response to inorganic substances using an online DAVID gene ontological analysis [54]. Several enzymes in metabolic processes are usually abundant and easily found on the 2-DE, whereas low-abundant proteins involve in gene regulation and signal transduction cascades are difficult to detect. However, 1D-SDS-PAGE may be used as a standard separation method for complex protein mixtures based on their molecular weight prior MS analysis. Gel-enhanced liquid chromatography coupled with tandem mass spectrometry (GeLC-MS/MS) is a direct intensive methodology that provides high coverage of the proteome. GeLC-MS/MS is applicable to reduce gel-to-gel variation and inconsistency of protein spots in a 2-DE and to increase sensitivity of detection of low-abundant proteins [55]. Current study of bone marrow plasma proteome profile using GeLC-MS/MS technique has reported differentially expressed proteins between patients and normal donors and showed that many of which have functions in intracellular signaling and metabolic pathways associated with ineffective erythropoiesis [51]. GeLC-MS/MS has been used to analyze the $CD34^{+}$/HSCs' thalassemic proteome during the in vitro differentiation of erythroid lineages and identified dysregulation of many signaling protein of differentiation processes in thalassemia patients (unpublished data).

4.2 LC-based technique

LC-based technique or shotgun proteomic analysis is an advanced and widely used approach for protein identification and characterization of sequence variants and posttranslational modifications [56]. Shotgun proteomics is considered as a bottom-up approach, by which short peptides are prepared by proteolytic digestion, usually by trypsin, and subjected to protein sequencing via tandem mass spectrometry (MS/MS) and automated protein database searching. After proteolytic digestion by trypsin, short peptide mixture is fractionated by hydrophobicity and charged using multidimensional liquid chromatography. Eluted peptides from the column are then ionized and separated by m/z in the first stage of MS/MS. Fragmentation is initiated when the peptide ions undergo collision-induced dissociation (CID) within inert gas. The ionized fragments are then separated by m/z in the second stage of MS/MS, and the corresponding peptide ions generate fragment ion spectral fingerprints for peptide identification. The process continues until all peptides are eluted from the chromatography column. Peptide identification is accomplished by comparing the fragmentation spectra derived from peptide fragmentation with theoretical MS/MS spectra based on in silico protein database. Protein inference is achieved by allocating peptide sequences to proteins [56–58]. Shotgun proteomics has a relatively high throughput compared to other MS-based proteomic technologies. This gel-free approach involves proteolytic digestion of proteins into short peptides (1–3 kDa) as they are easier to fragment than intact proteins simplifying MS/MS sequencing [21].

Erythropoiesis involves posttranslational modifications of proteins associated with cell-cell communications and cell interactions inside the bone marrow. Phosphorylation is mediated by kinase enzymes which have different target specificities. Different phosphorylation sites on the same protein usually serve different protein activities. Shotgun proteomic strategy becomes the widely used for studying phosphoproteomes. The CD34$^+$/HSCs' isolation from the bone marrow is a good representative model for elucidation signaling proteins using phosphoproteomic analysis comparing with thalassemia patient and normal donor. The workflow includes an enrichment of phosphoproteins among patient and donor samples using immobilized metal affinity chromatography (IMAC), followed by ESI-LC-MS/MS. This approach detects upregulation of signaling proteins of apoptosis pathway linked to ineffective erythropoiesis in β-thalassemic CD34$^+$/HSCs [15].

5. Alteration of proteome profiles in hematopoietic stem cells (HSCs) derived from β-thalassemia/HbE patients

The hallmarks of stem cells include self-renewal and differentiation ability to give rise different types of mature cells. Stem cell biology is an important field of biomedical science and holds promise in clinical applications. The hematopoietic stem cells (HSCs) are multipotent progenitor cells located in adult bone marrow, responsible for blood cell production. Multipotency or multi-lineage differentiation potential of the HSCs defines as the capacity to give rise to several differentiated cell types including myeloid progenitor cells, erythroid-megakaryocyte progenitor cells, and lymphoid progenitor cells [13]. In addition, the proliferation and differentiation of HSCs into the erythroid lineage are the acquisition of functional and structural properties associated with erythrocyte physiology [37, 59]. The cell isolation technique of the primitive HSCs' population expressing specific cell surface markers, CD34 and CD133, is a gold standard method widely used for investigating of erythropoiesis in vivo [60]. Using antibody-conjugated magnetic particles is the first step after the isolation of mononuclear cells from bone marrow aspirate or from peripheral blood. Optimized culture condition for in vivo erythroid development requires defined medium supplemented with cytokines, such as erythropoietin (EPO), to modulate erythroid cell development from HSCs/CD34$^+$. Erythroid differentiation is a multistep process starting when a subset of HSCs/CD34$^+$ commits to differentiate to the megakaryocyte-erythrocyte progenitors (MEPs). The differentiation of MEPs into the most immature erythroid progenitors, namely, a burst-forming unit erythroid (BFU-Es), involves a number of transcriptional regulators. The BFU-Es generate colony-forming units-erythroid cells (CFU-Es) which strictly depends on the erythropoietin to induce differentiation via erythropoietin receptor (EPOR). Subsequently, the CFU-Es develop into proerythroblasts (the first erythroid precursors) through the basophilic, the polychromatophilic, and the orthochromatic stages. During this differentiation process, the hemoglobin synthesis, cellular content modification, and nuclear condensation gradually take place, followed by the enucleation step generating reticulocytes which are immature red blood cells (RBCs). In addition to the macrophage-induced cellular death pathways, autophagy is the key mechanism of the biconcave formation of mature RBCs [13]. Conversely, in β-thalassemia/HbE patients, the ineffective erythropoiesis has dramatic effects in the erythroid lineage development, resulting in hyperproliferation, increased apoptosis, and inhibited terminal erythroid differentiation. To study the pathogenesis of β-thalassemia/HbE, our strategies include the isolation of $CD34^+$/HSCs, in vitro erythroid differentiation, and bone marrow plasma enrichment followed by the MS-based analysis to comprehensively understand the early molecular basis of β-thalassemia/HbE. This method revealed the bona fide proteome alterations of ineffective erythropoiesis from $CD34^+$/HSCs, the dynamic differentiation of $CD34^+$/HSCs in the erythroid lineage, and bone marrow microenvironment (**Figure 1A**). The bioinformatic approach has also been applied to analyze integrated thalassemia-specific proteomes in a range of patient specimens. We report interesting imbalance signaling landscapes which may contribute to ineffective erythropoiesis found in β-thalassemia/HbE patients.

5.1 Metabolic enzymes

Leecharoenkiat et al. reported the first differential 2-DE proteome between β^0-thalassemia/HbE and control erythroblasts isolated from peripheral $CD34^+$/HSCs. Many differentially expressed proteins identified in this study were constituents of the glycolysis and TCA pathways and were correlated with the degree of erythroid expansion [18]. Our investigation of bone marrow microenvironment proteome has shown increased level of ATP citrate lyase (ACLY) in β-thalassemia patients. ACLY is primarily responsible for synthesis of acetyl CoA in TCA cycle. Consequentially, hyperproliferation of erythroblast during ineffective erythropoiesis might be related with the increase of ACLY indicating a high metabolic flux governed by acetyl-CoA production. ACLY can support sufficient energy to drive the TCA cycle and mitochondrial oxidative phosphorylation [51]. Therefore, the comparative proteome profiling of peripheral blood $CD34^+$/HSCs and bone marrow plasma proteome indicated the similar target metabolic protein biomarkers associated with ineffective erythropoiesis.

5.2 Oxidative damage and antioxidant proteins

An imbalance between oxidative damage and antioxidant enzymes is associated with cellular pathology inside HSCs and differentiated erythroid precursors in β-thalassemia patients. Comparative proteome analysis during differentiation of $CD34^+$/HSCs through erythroid lineage for 7 days in β-thalassemia patients compared with normal showed many proteins differentially expressed; these include superoxide dismutase (SOD), peroxiredoxin 6 (PXD6), and peroxiredoxin 2 (PXD2). An increased level of PXD2 was found in differentiated thalassemic $CD34^+$/HSCs at day 0 through day 7 (unpublished data). In addition, PXD2 has function in stress response in the protective system during stress erythropoiesis in thalassemic mice model [61]. PXD2 is an abundant protein mostly identified in many proteomic studies from various clinical specimens of β-thalassemia [50, 54, 62]. An increased oxidative damage from alpha excess and iron overload thus triggers the cellular defensive mechanism, including antioxidative responses. Nevertheless, thalassemic stem cells could not prolong survival pathways to prevent cell death. Moreover, comparative bone marrow plasma proteome from patients and normal showed a decreased level of an antioxidant secretory or extracellular exosomal protein, namely, apolipoprotein D (ApoD), in patients who have an impairment of bone marrow microenvironment to dyserythropoietic condition [51].

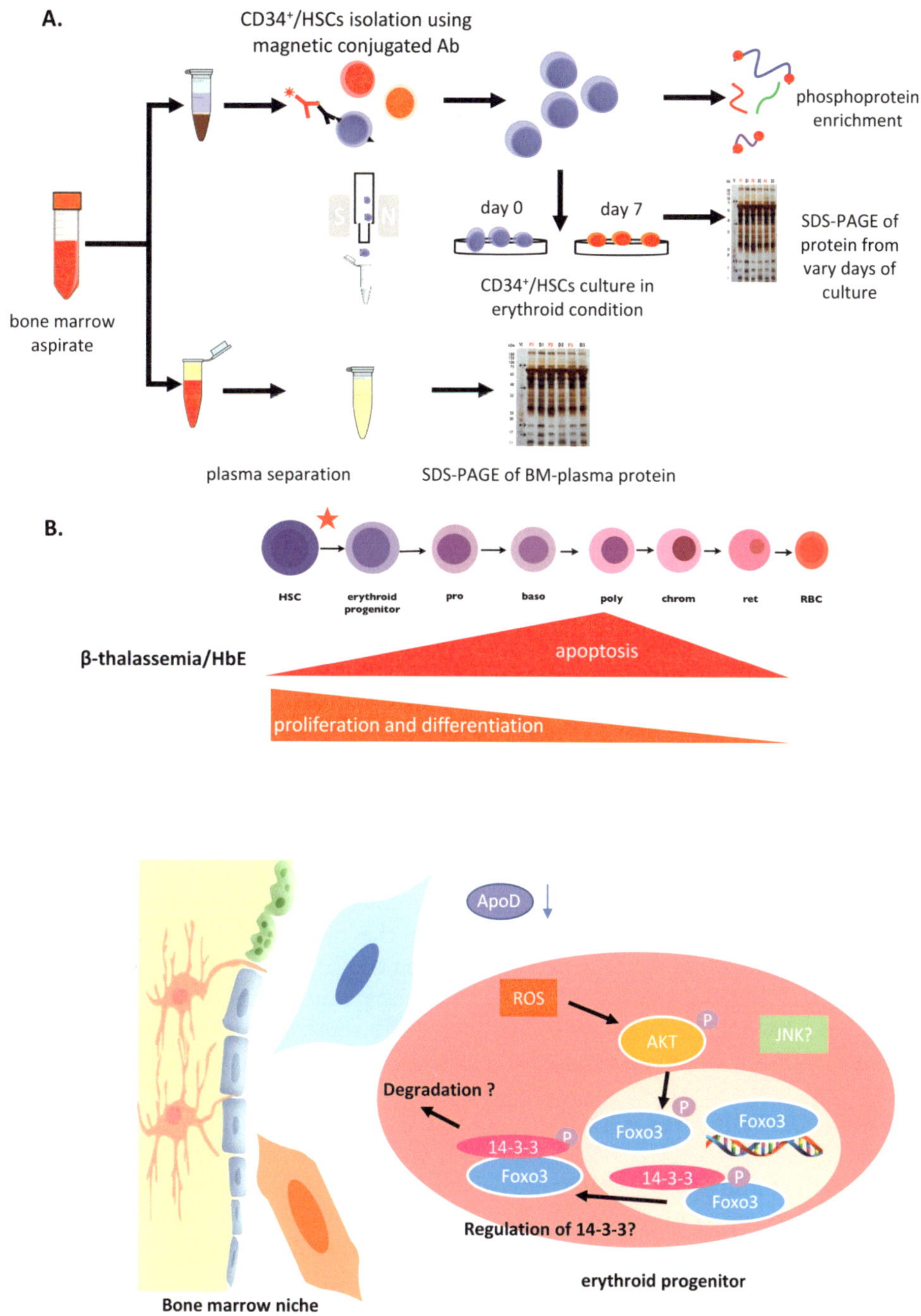

Figure 1.
Diagram showed the sample preparation from bone marrow-derived CD34⁺/HSCs and bone marrow plasma prior to MS-based proteomic approaches. The three sources of proteomes for investigating ineffective erythropoiesis include CD34⁺/HSCs' phosphoproteome, dynamic differential proteome during erythroid development, and bone marrow plasma proteome (A). Schematic diagram of identified proteins and proposed signaling pathway for ineffective erythropoiesis in β-thalassemia/HbE (B).

5.3 Heat shock protein

Recent studies have suggested that the molecular chaperone or the heat shock protein maintains cellular homeostasis during erythropoiesis. Regulated by the GATA-1, the master transcriptional machinery of erythropoiesis, the heat shock protein 70 (HSP70) plays an important role in protecting from caspase-3-mediated GATA-1 degradation [63]. Conversely, in β-thalassemia major, the alleviated free alpha globin exacerbates the oxidative damage inside erythroblast which induces the translocation of heat shock protein 70 from the nucleus to cytoplasm, leaving GATA1 unprotected [63]. The expression of heat shock protein 70 in patient group was dramatically significantly increased from day 0 to day 7, suggesting it as a contributing factor of ineffective erythropoiesis (unpublished data). Furthermore, the heat shock protein 90 that has a similar expression pattern as heat shock protein 70 and has been identified over expression from platelet-free plasma-derived microparticles proteome in β-thalassemia/HbE patients [50].

5.4 Apoptotic pathway

Erythroid progenitors undergoing terminal erythroid differentiation to generate sufficient erythroblasts and reticulocytes are impaired during ineffective erythropoiesis of β-thalassemia involving the acceleration of apoptotic cell death in erythroid precursors. The apoptotic proteins of intrinsic and extrinsic pathways have been firstly identified from the phosphoproteomic analysis of β-thalassemia $CD34^+$/HSCs. This might explain why even freshly isolated HSCs of β-thalassemia bone marrow showed less survival compared to HSCs from normal donors. Phosphoproteome analysis of β-thalassemia HSCs identified an increased expression of cytochrome C (CytC), apoptosis-inducing factor (AIF), and caspase 6 (CASP6) [15]. Moreover, extrinsic apoptotic pathway plays important roles during erythroid differentiation rather than the progenitor or HSCs' stages. The analysis of thalassemia HSCs' phosphoproteome during in vitro differentiation to erythroid lineage of $CD34^+$/HSCs revealed very high expression of death receptor proteins, such as TNF receptor-associated factor 2 (TRAF2), at day 7 but was undetectable at day 0 (unpublished data), as well as the downregulations of tumor necrosis factor ligand superfamily member 6 (FASL) and tumor necrosis factor receptor superfamily member 12A [15]. Thus, we propose that the imbalance of death signaling pathway could have a predominant effect during erythroid proliferation and differentiation in β-thalassemia. However, few studies explained how ROS and oxidative stress are related to an ineffective erythropoiesis. It remains to investigate whether ROS signaling could provoke apoptotic pathway of ineffective erythropoiesis.

5.5 The AKT/FOXO3/14-3-3 axis

After the finding of the imbalance cell death protein in $CD34^+$/HSCs' proteomes from both fresh isolated and erythroid differentiated condition *in vitro*. The 14-3-3 protein was found upregulated in HSCs of β-thalassemia/HbE patients. This identified a molecular linkage between both survival and death signaling pathways [15]. An elevated ROS level in β-thalassemia erythroid cells plays an important role in the activation of AKT, potentially resulting in the repression of FOXO3 activity and reducing the cellular responses to oxidative stress. Nevertheless, AKT-mediated FOXO3 phosphorylation at Ser253 could not maintain the activation of FOXO3 activity to induce downstream gene expression of stress responses in β-thalassemia/HbE [14]. This leads to a premature apoptosis of erythroid cells. In this circumstance, activation of AKT and the translocation of FOXO3 from the nucleus to cytoplasm involve the activation of 14-3-3 which binds to FOXO3 and induces FOXO3 degradation by ubiquitination/proteasome pathway [64]. In addition to the signaling crosstalk of AKT and 14-3-3 modulated FOXO3 function, c-Jun N-terminal kinase (JNK) pathway and other posttranslational modifications, especially acetylation, promote FOXO3 transcriptional activity inside the nucleus [64, 65]. However, the regulation of 14-3-3 among AKT and JNK signaling-mediated FOXO3 function through involving imbalance of death and survival signaling during ineffective erythropoiesis remains to be investigated (**Figure 1B**).

6. Conclusions

Proteomic analysis by MS has become an efficient tool for investigating pathophysiology of β-thalassemia/HbE overcoming limiting factors of stem cell samples, especially hematopoietic stem cells (HSCs). Integrated proteome profiling using shotgun-based and gel-based proteomic analyses of clinical bone marrow or peripheral blood samples shed light on the molecular mechanisms on ineffective erythropoiesis in β-thalassemia. Better understanding of these molecular mechanisms will help the development of novel treatment of the disease.

Acknowledgements

The authors gratefully acknowledge the financial support provided by Thammasat University under the TU New Research Scholar, Contract No. 1/2015, and National Research Council of Thailand, Contract Nos. 3949 and 43000.

Conflict of interest

The authors have no potential conflict of interest to disclose.

Author details

Saranyoo Ponnikorn[1*], Siripath Peter Kong[1], Sasipim Thitivirachawat[1], Chanawin Tanjasiri[1], Sumalee Tungpradabkul[2] and Suradej Hongeng[3]

1 Chulabhorn International College of Medicine, Thammasat University Rangsit Campus, Pathum Thani, Thailand

2 Department of Biochemistry, Faculty of Science, Mahidol University, Bangkok, Thailand

3 Department of Pediatrics, Faculty of Medicine Ramathibodi Hospital, Mahidol University, Bangkok, Thailand

*Address all correspondence to: saranyoo@tu.ac.th

References

[1] Thein SL, Menzel S, Lathrop M, Garner C. Control of fetal hemoglobin: New insights emerging from genomics and clinical implications. Human Molecular Genetics. 2009;**18**(R2):R216-R223

[2] Thein SL. Genetic modifiers of the beta-haemoglobinopathies. British Journal of Haematology. 2008;**141**(3):357-366

[3] Fucharoen S, Weatherall DJ. The hemoglobin E thalassemias. Cold Spring Harbor Perspectives in Medicine. 2012;**2**:8

[4] Weatherall DJ. Thalassemia as a global health problem: Recent progress toward its control in the developing countries. Annals of the New York Academy of Sciences. 2010;**1202**:17-23

[5] Thein SL. Genetic association studies in beta-hemoglobinopathies. Hematology American Society of Hematology Education Program. 2013;**2013**:354-361

[6] Riewpaiboon A, Nuchprayoon I, Torcharus K, Indaratna K, Thavorncharoensap M, Ubol BO. Economic burden of beta-thalassemia/Hb E and beta-thalassemia major in Thai children. BMC Research Notes. 2010;**3**:29

[7] Rund D. Thalassemia 2016: Modern medicine battles an ancient disease. American Journal of Hematology. 2016;**91**(1):15-21

[8] Rund D, Rachmilewitz E. Beta-thalassemia. The New England Journal of Medicine. 2005;**353**(11):1135-1146

[9] Hirsch RE, Sibmooh N, Fucharoen S, Friedman JM. HbE/beta-thalassemia and oxidative stress: The key to pathophysiological mechanisms and novel therapeutics. Antioxidants & Redox Signaling. 2017;**26**(14):794-813

[10] Pootrakul P, Sirankapracha P, Hemsorach S, Moungsub W, Kumbunlue R, Piangitjagum A, et al. A correlation of erythrokinetics, ineffective erythropoiesis, and erythroid precursor apoptosis in Thai patients with thalassemia. Blood. 2000;**96**(7):2606-2612

[11] Ribeil JA, Arlet JB, Dussiot M, Moura IC, Courtois G, Hermine O. Ineffective erythropoiesis in beta-thalassemia. Scientific World Journal. 2013;**2013**:394295

[12] Koulnis M, Porpiglia E, Hidalgo D, Socolovsky M. Erythropoiesis: From molecular pathways to system properties. Advances in Experimental Medicine and Biology. 2014;**844**:37-58

[13] Nandakumar SK, Ulirsch JC, Sankaran VG. Advances in understanding erythropoiesis: Evolving perspectives. British Journal of Haematology. 2016;**173**(2):206-218

[14] Thanuthanakhun N, Nuntakarn L, Sampattavanich S, Anurathapan U, Phuphanitcharoenkun S, Pornpaiboonstid S, et al. Investigation of FoxO3 dynamics during erythroblast development in beta-thalassemia major. PLoS One. 2017;**12**(11):e0187610

[15] Ponnikorn S, Panichakul T, Sresanga K, Wongborisuth C, Roytrakul S, Hongeng S, et al. Phosphoproteomic analysis of apoptotic hematopoietic stem cells from hemoglobin E/beta-thalassemia. Journal of Translational Medicine. 2011;**9**:96

[16] Khungwanmaythawee K, Sornjai W, Paemanee A, Jaratsittisin J, Fucharoen S, Svasti S, et al. Mitochondrial changes in beta0-thalassemia/Hb E disease. PLoS One. 2016;**11**(4):e0153831

[17] Lithanatudom P, Wannatung T, Leecharoenkiat A, Svasti S,

Fucharoen S, Smith DR. Enhanced activation of autophagy in beta-thalassemia/Hb E erythroblasts during erythropoiesis. Annals of Hematology. 2011;**90**(7):747-758

[18] Leecharoenkiat A, Wannatung T, Lithanatudom P, Svasti S, Fucharoen S, Chokchaichamnankit D, et al. Increased oxidative metabolism is associated with erythroid precursor expansion in beta0-thalassaemia/Hb E disease. Blood Cells, Molecules & Diseases. 2011;**47**(3):143-157

[19] Lithanatudom P, Leecharoenkiat A, Wannatung T, Svasti S, Fucharoen S, Smith DR. A mechanism of ineffective erythropoiesis in beta-thalassemia/Hb E disease. Haematologica. 2010;**95**(5):716-723

[20] Wannatung T, Lithanatudom P, Leecharoenkiat A, Svasti S, Fucharoen S, Smith DR. Increased erythropoiesis of beta-thalassaemia/Hb E proerythroblasts is mediated by high basal levels of ERK1/2 activation. British Journal of Haematology. 2009;**146**(5):557-568

[21] Zhang Y, Fonslow BR, Shan B, Baek M-C, Yates JR 3rd. Protein analysis by shotgun/bottom-up proteomics. Chemical Reviews. 2013;**113**(4):2343-2394

[22] Baggerman G, Vierstraete E, De Loof A, Schoofs L. Gel-based versus gel-free proteomics: A review. Combinatorial Chemistry & High Throughput Screening. 2005;**8**(8):669-677

[23] Weatherall D. Thalassemia: The long road from the bedside through the laboratory to the community. Nature Medicine. 2010;**16**(10):1112-1115

[24] Weatherall D, Akinyanju O, Fucharoen S, et al. Inherited disorders of hemoglobin. In: Jamison DT, Breman JG, Measham AR, et al., editors. Disease Control Priorities in Developing Countries. 2nd edition. Washington (DC): The International Bank for Reconstruction and Development/The World Bank; 2006. Chapter 34

[25] Weatherall DJ. The definition and epidemiology of non-transfusion-dependent thalassemia. Blood Reviews. 2012;**26**(Suppl 1):S3-S6

[26] Clark BE, Thein SL. Molecular diagnosis of haemoglobin disorders. Clinical and Laboratory Haematology. 2004;**26**(3):159-176

[27] Olivieri NF, Thayalsuthan V, O'Donnell A, Premawardhena A, Rigobon C, Muraca G, et al. Emerging insights in the management of hemoglobin E beta thalassemia. Annals of the New York Academy of Sciences. 2010;**1202**:155-157

[28] Winichagoon P, Fucharoen S, Chen P, Wasi P. Genetic factors affecting clinical severity in beta-thalassemia syndromes. Journal of Pediatric Hematology/Oncology. 2000;**22**(6):573-580

[29] Fucharoen S, Ketvichit P, Pootrakul P, Siritanaratkul N, Piankijagum A, Wasi P. Clinical manifestation of beta-thalassemia/hemoglobin E disease. Journal of Pediatric Hematology/Oncology. 2000;**22**(6):552-557

[30] Fucharoen S, Winichagoon P. Clinical and hematologic aspects of hemoglobin E beta-thalassemia. Current Opinion in Hematology. 2000;7(2):106-112

[31] Thein SL. Pathophysiology of beta thalassemia—A guide to molecular therapies. Hematology American Society of Hematology Education Program. 2005;**1**:31-37

[32] Gardenghi S, Grady RW, Rivella S. Anemia, ineffective erythropoiesis,

and hepcidin: Interacting factors in abnormal iron metabolism leading to iron overload in beta-thalassemia. Hematology/Oncology Clinics of North America. 2010;**24**(6):1089-1107

[33] Melchiori L, Gardenghi S, Rivella S. beta-Thalassemia: HiJAKing ineffective erythropoiesis and iron overload. Advances in Hematology. 2010;**2010**:938640

[34] Mathias LA, Fisher TC, Zeng L, Meiselman HJ, Weinberg KI, Hiti AL, et al. Ineffective erythropoiesis in beta-thalassemia major is due to apoptosis at the polychromatophilic normoblast stage. Experimental Hematology. 2000;**28**(12):1343-1353

[35] Centis F, Tabellini L, Lucarelli G, Buffi O, Tonucci P, Persini B, et al. The importance of erythroid expansion in determining the extent of apoptosis in erythroid precursors in patients with beta-thalassemia major. Blood. 2000;**96**(10):3624-3629

[36] Yuan J, Angelucci E, Lucarelli G, Aljurf M, Snyder LM, Kiefer CR, et al. Accelerated programmed cell death (apoptosis) in erythroid precursors of patients with severe beta-thalassemia (Cooley's anemia). Blood. 1993;**82**(2):374-377

[37] Liang R, Ghaffari S. Advances in understanding the mechanisms of erythropoiesis in homeostasis and disease. British Journal of Haematology. 2016;**174**(5):661-673

[38] Libani IV, Guy EC, Melchiori L, Schiro R, Ramos P, Breda L, et al. Decreased differentiation of erythroid cells exacerbates ineffective erythropoiesis in beta-thalassemia. Blood. 2008;**112**(3):875-885

[39] Camaschella C, Pagani A, Nai A, Silvestri L. The mutual control of iron and erythropoiesis. International Journal of Laboratory Hematology. 2016;**38**(Suppl 1):20-26

[40] Rasool M, Malik A, Jabbar U, Begum I, Qazi MH, Asif M, et al. Effect of iron overload on renal functions and oxidative stress in beta thalassemia patients. Saudi Medical Journal. 2016;**37**(11):1239-1242

[41] Paluru P, Hudock KM, Cheng X, Mills JA, Ying L, Galvao AM, et al. The negative impact of Wnt signaling on megakaryocyte and primitive erythroid progenitors derived from human embryonic stem cells. Stem Cell Research. 2014;**12**(2):441-451

[42] Steinert R, von Hoegen P, Fels LM, Gunther K, Lippert H, Reymond MA. Proteomic prediction of disease outcome in cancer: Clinical framework and current status. American Journal of Pharmacogenomics. 2003;**3**(2):107-115

[43] Choudhary C, Mann M. Decoding signalling networks by mass spectrometry-based proteomics. Nature Reviews Molecular Cell Biology. 2010;**11**(6):427-439

[44] Hanash S. The emerging field of protein microarrays. Proteomics. 2003;**3**(11):2075

[45] Hanash S. Disease proteomics. Nature. 2003;**422**(6928):226-232

[46] Pagel O, Loroch S, Sickmann A, Zahedi RP. Current strategies and findings in clinically relevant post-translational modification-specific proteomics. Expert Review of Proteomics. 2015;**12**(3):235-253

[47] Liumbruno G, D'Alessandro A, Grazzini G, Zolla L. How has proteomics informed transfusion biology so far? Critical Reviews in Oncology/Hematology. 2010;**76**(3):153-172

[48] D'Alessandro A, Zolla L. Proteomic analysis of red blood cells and the potential for the clinic: What have we learned so far? Expert Review of Proteomics. 2017;**14**(3):243-252

[49] Kittivorapart J, Crew VK, Wilson MC, Heesom KJ, Siritanaratkul N, Toye AM. Quantitative proteomics of plasma vesicles identify novel biomarkers for hemoglobin E/beta-thalassemic patients. Blood Advance. 2018;**2**(2):95-104

[50] Chaichompoo P, Kumya P, Khowawisetsut L, Chiangjong W, Chaiyarit S, Pongsakul N, et al. Characterizations and proteome analysis of platelet-free plasma-derived microparticles in beta-thalassemia/hemoglobin E patients. Journal of Proteomics. 2012;**76**(Spec):239-250

[51] Ponnikorn S, Mongkolrob R, Klongthalay S, Roytrakul S, Srisanga K, Tungpradabkul S, et al. Comparative proteome-wide analysis of bone marrow microenvironment of beta-thalassemia/hemoglobin E. Proteome. 2019;**7**(1)

[52] Chevalier F. Highlights on the capacities of "gel-based" proteomics. Proteome Science. 2010;**8**:23

[53] Kim YI, Cho JY. Gel-based proteomics in disease research: Is it still valuable? Biochimica et Biophysica Acta, Proteins and Proteomics. 2019;**1867**(1):9-16

[54] Lithanatudom P, Smith DR. Analysis of protein profiling studies of beta-thalassemia/Hb E disease. Proteomics Clinical Applications. 2016;**10**(11):1093-1102

[55] Paemanee A, Wikan N, Roytrakul S, Smith DR. Application of GelC-MS/MS to proteomic profiling of chikungunya virus infection: Preparation of peptides for analysis. Methods in Molecular Biology. 2016;**1426**:179-193

[56] Zhou H, Ning Z, Starr AE, Abu-Farha M, Figeys D. Advancements in top-down proteomics. Analytical Chemistry. 2012;**84**(2):720-734

[57] Hung CW, Tholey A. Tandem mass tag protein labeling for top-down identification and quantification. Analytical Chemistry. 2012;**84**(1):161-170

[58] Liu X, Sirotkin Y, Shen Y, Anderson G, Tsai YS, Ting YS, et al. Protein identification using top-down. Molecular & Cellular Proteomics. 2012;**11**(6):M111 008524

[59] Cumano A, Godin I. Ontogeny of the hematopoietic system. Annual Review of Immunology. 2007;**25**:745-785

[60] Panichakul T, Payuhakrit W, Panburana P, Wongborisuth C, Hongeng S, Udomsangpetch R. Suppression of erythroid development in vitro by plasmodium vivax. Malaria Journal. 2012;**11**:173

[61] Matte A, Low PS, Turrini F, Bertoldi M, Campanella ME, Spano D, et al. Peroxiredoxin-2 expression is increased in beta-thalassemic mouse red cells but is displaced from the membrane as a marker of oxidative stress. Free Radical Biology & Medicine. 2010;**49**(3):457-466

[62] Bhattacharya D, Saha S, Basu S, Chakravarty S, Chakravarty A, Banerjee D, et al. Differential regulation of redox proteins and chaperones in HbEbeta-thalassemia erythrocyte proteome. Proteomics Clinical Applications. 2010;**4**(5):480-488

[63] Arlet JB, Ribeil JA, Guillem F, Negre O, Hazoume A, Marcion G, et al. HSP70 sequestration by free alpha-globin promotes ineffective erythropoiesis in beta-thalassaemia. Nature. 2014;**514**(7521):242-246

[64] Huang H, Tindall DJ. Dynamic FoxO transcription factors. Journal of Cell Science. 2007;**120**(Pt 15):2479-2487

[65] Maiese K, Chong ZZ, Hou J, Shang YC. The "O" class: Crafting clinical care with FoxO transcription factors. Advances in Experimental Medicine and Biology. 2009;**665**:242-260

Chapter 5

Mass Spectrometry for Cancer Biomarkers

Radu Albulescu, Andrei Jose Petrescu, Mirela Sarbu, Alice Grigore, Raluca Ica, Cristian V. A. Munteanu, Adrian Albulescu, Ioana V. Militaru, Alina-Diana Zamfir, Stefana Petrescu and Cristiana Tanase

Abstract

Cancer is considered as the second cause of morbidity and also mortality, after cardiovascular diseases. Despite the immense progress in efficacious biomarkers made in the present time, there are very few of them that can timely detect cancers or that can predict treatment outcomes or stratify patients according to their response to the treatment. Among other modern instruments involved in the research of this disease, proteomics emerged strongly, since it analyzes the "molecular effectors." Although it has some setbacks (like the lack of amplification of the signal), it is however one of the best means of investigating the presence and causes and predicting the evolution patterns of the disease. This chapter describes briefly pre-analytical (pre-MS steps), the main concepts, and the MS equipment used for such applications, followed by the presentation of several proteomic applications in melanoma, glioblastoma, pancreatic cancer, and colon cancer.

Keywords: proteomics, biomarkers, melanoma, glioblastoma, pancreatic cancer, colorectal cancer

1. Introduction

Cancer is a major challenge for health, representing the second leading cause of death worldwide. While certain progress was made in diagnostics and in some therapies for different cancers, there is still an urgent need for consistent efforts in order to set up new biomarkers for the diagnostics, prognosis, or stratification of patients. Efforts have been made to identify potential biomarkers by different approaches, like the use of techniques of genomics, transcriptomics, metabolomics, or proteomics. While the nucleic acid-based techniques present the advantage of the amplification potential, but lack the "effector" role, proteomics is considered more relevant, since proteome is directly involved in producing effective biological effects, yet it lacks the potential of amplification of nucleic acids. So, the detection of modifications, quantitative and/or qualitative, mainly affecting, for instance, low abundance proteins in biological fluids, requires increased sensitivity equipment and eventual strategies to eliminate the high-abundant and less significant ones. All these represent a major challenge of today's proteomics and will be briefly reviewed.

2. Preliminary steps

One major concern in proteomics is sample preparation, since it is generally accepted that this step is generating most of the false discoveries in the field. Therefore, we are briefly referring to this step, as well as to pre-MS steps, which are dedicated to reducing the sample complexity. Even if we do not address a total proteome, the analysis of a plasma proteome or a cellular proteome generates a large amount of data. Thus, different strategies of pre-separation are used in the proteomic analysis.

Depending on the biological materials used as start materials, it is possible that additional depletion technologies are needed, in order to reduce/eliminate major proteins (such as serum albumin or immunoglobulins).

One of the best technologies established uses an electrophoretic separation step to separate proteins; in earlier years, even the old one-dimensional sodium dodecyl sulfate (1D SDS) gel electrophoresis was used for the purpose. The next step in this approach was the use of 2D electrophoresis, which is more efficient since it uses the combination of isoelectric focusing and mass separation of proteins. The critical aspect is the time and material consumption, since for each condition one needs several runs to provide statistically significant results. Finally, the more advanced two-dimensional fluorescence difference gel electrophoresis (2D-DIGE) was developed, which provides a significant reduction of runs (since on each gel three distinct conditions can be analyzed, such as sample, control, and sample-control mixture), and has about two orders of concentration lower. This last approach affords a good detection of low abundance proteins.

Another means of achieving the first separation of proteins is provided by capillary electrophoresis (CE). In combination with mass spectrometry (MS), this technique allows the detection of proteins [1], as well as glycoproteins [2]; besides proteomics, the technology provides in combination with MS solid platform for metabolomics studies [3, 4].

More recently, a distinct strategy was adopted, consisting in the proteolytic fragmentation of the proteins in samples (i.e., using trypsin), followed by the isoelectric focusing of the resulting peptide samples. The process usually uses single-use devices that are based on very narrow pH steps for focusing. Usually, such devices serve as the "feeder" for the mass spectrometry device, and one can obtain a complete picture of the proteins present in the sample.

3. Mass spectrometers for proteomics

Over the past decades, mass spectrometry-based proteomics was developed as a fundamental tool in identifying clinically relevant biomarkers for disease progression and assessing the organism response to treatment, with better outcome for patients.

Generally, proteomics refers to any kind of large-scale characterization of protein species in a sample given by an analytical approach, yet due to its extensive use, proteomics becomes quite synonymous with mass spectrometry analysis for protein identification, localization, interaction, abundance, and posttranslational modifications lately [5]. More specifically, in terms of biomarker discovery, MS proteomics usually indicates proteome characterization and comparison among various disease-stage models and healthy subjects.

Compared with the genomics or transcriptomics technologies, proteomics remained for years limited in terms of throughput and depth [6]. This is due to the fact that unlike nucleic acids, proteins cannot be easily amplified which limits the detection of the least abundant species in complex biological samples.

Another challenge is due to protein posttranslational modifications (PTMs). PTMs increase the complexity and dynamic range of proteome samples, compared to the genome samples, by creating a diversification that reach millions of modified protein products given the existence of protein isoforms, alternative splicing, somatic DNA rearrangements, or proteolytic events [5].

Although recent technological improvements in instrumentation also allow the analysis of intact proteins and protein complexes [6–9], there is still limited progress in adopting top-down MS strategy in a large-scale analysis of proteomes. As a consequence, the bottom-up strategy remains the main approach in MS-based proteomics [10]. In bottom-up workflows, proteins are first isolated from the biological sample, digested using single or multiple proteases to peptides (depending on the aim of the experiment), and then analyzed by mass spectrometry [11].

Due to proteome complexity, most often samples are subject to online separation in nano- or capillary C18 reversed-phase systems and even for higher dimensionality to offline or online fractionation using different chemistries at the protein or peptide level [12].

Traditionally, ion traps have been employed in discovery studies, in contrast to validation steps where triple quadrupole-type instruments have been used for targeted quantification in a larger number of samples [13].

Ion traps can store ions by applying multiple electrical fields from different directions either effectively "catching" the ions for further fragmentation or ejecting them for detection [14].

Nowadays, electrospray ionization (ESI) MS constitutes one of the most powerful analytical techniques, being capable of differentiating the pattern, mostly of proteins, between tumor and normal adjacent tissues, thus facilitating the elucidation of human tumor molecular fingerprint [15] and understanding the molecular alterations that take place during pathological processes.

ESI is a modern ionization technique that brought a big advantage in mass spectrometry since it can ionize large nonvolatile molecules without denaturation, keeping non-covalent, receptor-ligand complexes intact. As a soft ionization method, which takes place at atmospheric pressure, ESI causes little or no fragmentation of the molecules. The principle of the ESI technique is relatively simple: the sample dissolved in a polar solvent and possibly with added electrolytes up to a concentration of approximately 0.001–10 mM and passes at a slow rate (0.1–30 μl/min) through a very thin capillary (10–100 μm) held at a high potential (2–5 kV) that has an open end with a sharply pointed tip. Several parameters, among them the solvent, capillary shape, desolvation gas pressure and composition, and distance between the capillary and the counterelectrode, were found to directly influence the ESI potential value. In ESI process, a fine spray of highly charged droplets is generated and directed toward the counterelectrode. The droplets, usually smaller than 10 μm across, contain both solvent and analyte molecules. Further, by passing the droplets through a heated inert bath gas, i.e., nitrogen, the desolvation process occurs up to the moment when the surface tension can no longer sustain the charge (the Rayleigh limit). Subsequently, a Coulomb explosion takes place (**Figure 1**), generating the droplet disruption concomitant with singly or multiply charged analyte ion formation [16].

In ESI, the charge state of an analyte depends on the molecular structure and also on the ionizability of the molecule in a given solvent, and hence on the solvent pH, concentration, and composition [17]. In the case of small molecules (molecular weight below 2 kDa) such as metabolites, drugs, amino acids, glycans, and short peptide chains, only singly to triply charged ions are formed, while for large molecules like polypeptides and proteins (molecular weight exceeding over 2–3 kDa), multi-charged ions are mostly generated in a positive ion mode. Therefore, ESI mass spectra display multiple signals associated with species with multiple charge

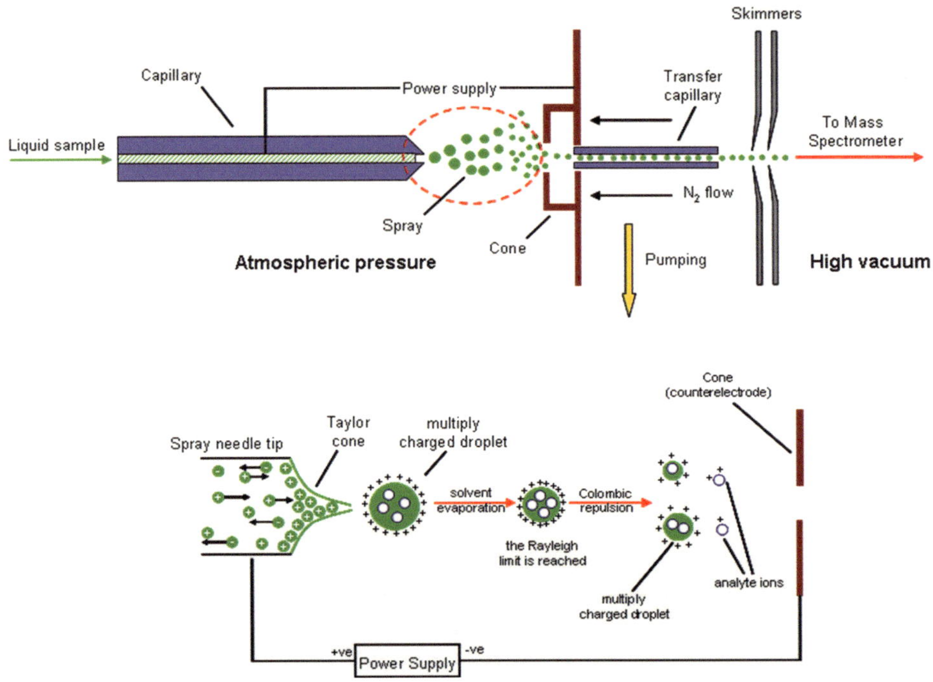

Figure 1.
The main physicochemical processes of electrospray ionization in positive ion mode.

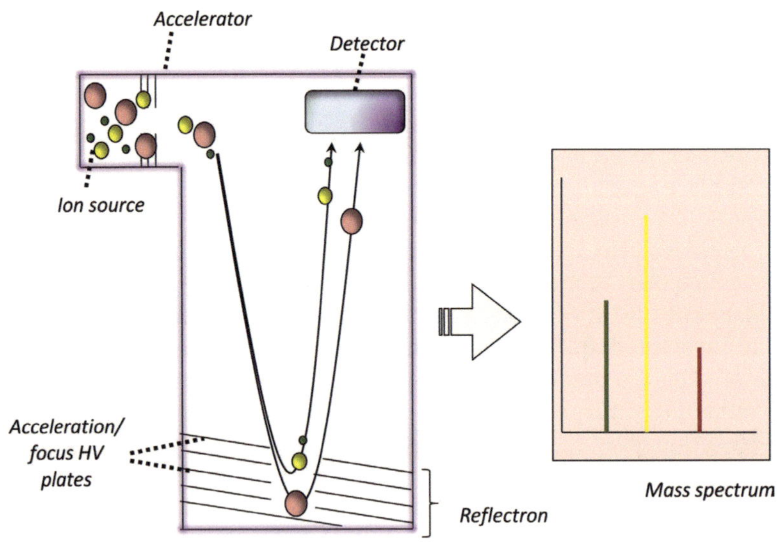

Figure 2.
Principles of ion separation in a time-of-flight analyzer.

states [17–19]. Although it brings complexity to the interpretation of the spectra, the generation of preponderantly multiply charged ions, either positive $[M + nH]^{n+}$ or negative $[M - nH]^{n-}$, represents a major advantage as it decreases the *m/z* ratios. Thus, larger molecules, i.e., biopolymers such as polypeptides, proteins, nucleic acids, anticancer drugs, etc., can be characterized by MS, using almost any type of analyzer with which ESI is compatible (**Figure 2**).

Besides ESI, one of the most frequently used ionization sources in cancer protein biomarker discovery is matrix-assisted laser desorption/ionization (MALDI).

Often considered harmonizing to its sister "soft ionization" technique ESI, MALDI is based on analyte molecule co-crystallization with an excess of matrix material (mainly small organic molecules, present in 100–100,000-fold excess compared with sample concentration). After the co-crystals were hit by pulsed laser beam (several pulses per nanoseconds), the matrix absorbs the energy of the laser; it releases the energy into the sample as heat, causing the sample to vaporize and form ions [20]. Typical MALDI devices are equipped with UV N_2 lasers with an emission wavelength of 337 nm.

The most challenging step when MALDI is used as ionization source is the selection of the optimal matrix. The matrix must fulfill a series of requirements, such as (i) to provide homogenous co-crystals with the analyte molecules, (ii) to be stable under high vacuum to prevent its sublimation, (iii) to deliver a high absorbance at the emission wavelength of the laser, (iv) to offer a high signal-to-noise ratio for the investigated sample (high sensitivity), and (v) to exhibit a low tendency of analyte-matrix ion cluster formation.

Unlike ESI, the molecules ionized by MALDI are singly charged, a feature which greatly simplifies the data analysis. Although vacuum MALDI has the capacity of a higher throughput and more tolerance for contaminants than ESI, the loss of labile moieties represents the major drawback [20–22]. However, MALDI technique is a widespread ionization technique employed in mass spectrometry imaging (MSI). MALDI MSI emerged over the last years as a key technology for label-free mapping of the spatial distribution of pharmaceuticals or biomolecules, ranging from small metabolites to large proteins, in tissue sections [23].

In MALDI MSI, a focused ionization beam is used to analyze a specific region of thin tissue slices mounted on conductive microscope slides by generating a mass spectrum that is stored along with the spatial coordination where the measurement took place. Further, by moving stepwise the sample or the ionization beam and full scan of the sample, several thousands of distribution maps, or ion images, are generated. By choosing a peak of interest from the combined spectra, the MS data maps its distribution across the sample, as a function of x and y locations, revealing therefore greater insight and aiding the understanding of molecular makeup and regional heterogeneity. Based on the high spatial resolution (>10 μm), fast acquisition speed, robustness, and relative ease handling, an entire mouse brain can be imaged successively in both positive and negative ion modes with $50 \times 50\ \mu m^2$ pixels in less than 1 h by using MALDI-TOF-MSI [24].

The remarkable ability of localizing the biomolecules in tissues, even without any previous information about them, and of differentiating compounds by molecular weight, all with no need of sample cleanup or chromatography step prior to ionization, exponentially expanded the area of MALDI MSI applicability. Therefore, the method had recent major contributions in understanding the disorders by tracking proteins, lipids, and cell metabolism for correlating the biomolecular changes with diseases and improving diagnostics and drug delivery [25].

Although low-resolution instruments such as ion traps can be used for proteome characterization due to their fragmentation flexibility and fast cycle rates [26], high-resolution mass spectrometry (HRMS) remains a goal and ambition in proteomics. This is because the coeluting ions can overlap when measuring peptides in complex biological samples [27], and this can preclude the accurate charge state evaluation and quantification [28].

Although, initially, HRMS was available for the expensive Fourier-transform ion cyclotron resonance (FT-ICR) and magnetic instruments and only time-of-flight (TOF) instruments could be routinely used in facilities [29], developments in the late 1990s of technological ion trapping and storage resulted in the design of ion orbital trapping—*Orbitrap*—which allowed HRMS to be attained [30].

The Orbitrap consists of two outer electrodes (disposed in a special bell-like architecture) and a central spindle-like electrode between which ions are squeezed using electrical fields [31]. The *m/z* values are obtained by the Fourier transform (FT) of the image current produced by the axial oscillations of the ions injected in the mass analyzer [6].

The initial limiting factor in introducing Orbitrap in the commercial market was related to the difficulties in interfacing this system with ion sources such as electrospray ionization [32], but adding a curved ion trap, subsequently termed C-trap, for ion injection and coupling linear ion traps (LTQ) with Orbitrap, resulted in the introduction of a hybrid instrument which offered a good balance between speed, sensitivity, and resolution [33], as most of the methods in proteomics used HRMS for precursor mass measurement and the ion trap for ion fragment scanning.

Using this setup, it was possible to obtain an almost complete proteome mapping for several simpler organisms such as *E. coli* or yeasts in just a few hours [34–36].

Although it did not provide a complete proteomic map for more complex organisms, this had a significant impact in the characterization of proteomic signatures in many eukaryotic cell lines such as melanoma cells.

Mass spectrometry was found to be effective also in proteomic phenotyping of melanoma cell lines [37]. For example, Pirmoradian et al. reported the identification of more than 5000 protein groups in A375 human melanoma cell line using 4-h-long optimized gradients without any prefractionation [38]. They were able to reach this depth of the proteome using a relatively recently commercially introduced instrument setup which coupled the Orbitrap mass analyzer with a quadrupole mass filter and an enhanced FT deconvolution algorithm [39].

Besides proteomics, the hybrid platform is also instrumental in the assessment of the dynamics of protein interactions in the pathological context of melanoma [40]. This is important, since it was shown that mutations in p53 oncogene could alter its interaction partners leading to protein complexes that promote invasion and migration of tumor cells [40–42].

4. Melanoma biomarkers

The introduction of the Orbitrap mass analyzer allowed routine HRMS to be also used for biomarker discovery studies. Using an LTQ-Orbitrap, Kawahara et al. defined the phenotypic differences in the secretome of melanoma, carcinoma, and noncancerous cells [43]. It is interesting to note that the top two candidates from the carcinoma secretome were validated in the saliva of patients using pseudo-selected reaction monitoring (pseudo-SRM) experiments on the same conventional hybrid instrument, usually dedicated only to the discovery approach, which subjects the flexibility and the impact of this platform.

Counting about 55,500 deaths annually, cutaneous melanoma represents the most aggressive type of skin cancer [44]. In principle, melanoma biomarkers can provide insights into either the overall outcome of this disease or to the therapeutic response to available treatments, being classified as prognostic biomarkers, predictive biomarkers, and on-treatment biomarkers [45]. However, to date a relatively low number of melanoma biomarkers were found, making it difficult to predict the outcome in patients. Moreover, their expression levels change with the stage of the disease [46]. Importantly, melanoma seems to have the highest mutational load, and once it appeared it becomes life-threatening [47]. Several mutations are frequently linked to cutaneous malignant melanoma, and these mutations are presented in more than 85% of all new diagnosed melanoma cases.

Stable isotope labeling with amino acids mass spectrometry (SILAC-MS)—involving incorporation of labeled amino acids into all cell proteins—was recently used in melanoma biomarker discovery. The main advantage of this method is its quantitative accuracy and reproducibility [1]. Using SILAC coupled with nano-spray tandem MS, Liu et al. highlighted several differentially expressed proteins in MA2 cells. It is the case of CUB-domain-containing protein 1 (CDCP1), a surface marker especially expressed on cells with high metastatic feature [48]. The group of Janostiak performed SILAC and tried to identify potential MELK targets. They identify MELK as a regulator of the NF-kB pathway with implications in tumor survival. MELK inhibition blocks melanoma growth; therefore, it can be viewed as a potential therapeutic target [49].

One of the most interesting approaches in cancer research is the possibility of performing proteomics on formalin-fixed paraffin-embedded (FFPE) tissues in order to denote disease-state-linked molecular fingerprints. This is significant given that clinical research units collect patient tissue and usually process them for histopathology analysis by FFPE which cross-link protein amino acids for long-term storage [50–53]. Using an LTQ-Orbitrap system and a spectral count-based quantitative strategy, [54] report the identification of a comprehensive list of 171 possible disease-state-linked melanoma biomarkers that were found as differently expressed proteins between the three analyzed stages of melanoma: benign nevi, primary, and metastatic melanoma. Quantitative MS analysis was also performed by Sengupta's group to study histone post-translational modifications in melanoma cells. Considering their findings EZH2 (histone methyltransferase) was upregulated in both FFPE samples and melanoma cell lines in concordance with increased histone H3 lysine 27 trimethylation [16]. Despite the potential to give a more appropriate insight into melanoma environment, this technique presents some limitations due to formalin-induced covalent linkage of proteins [55].

An even more exciting perspective of HRMS is the investigation of PTMs—as these can impact protein folding, traffic, degradation, signaling pathways, and ultimately cellular response [56]. This is of special interest given that these modifications cannot be captured by genomics or transcriptomics studies, while the biomarkers for cancer could be also affected by such modifications.

Using multienzyme digestion and different fragmentation techniques [57] available on the hybrid ion trap-Orbitrap platform, Chiritoiu et al. reported the occupancy of all N-glycosylation sites of tyrosinase in a melanoma cell line [58], an enzyme shown to be correlated with patient clinical evolution in melanoma [59, 60]. Moreover, recent evidence suggests that even other members of the tyrosinase-related protein family could be also linked to melanoma progression [61, 62].

An emerging field in biomarker discovery is MSI in which the analysis is performed directly on frozen tissue in order to evaluate possible disease-specific molecular fingerprints. It was already shown that MSI could be performed using both low- and high-resolution instruments [63–65] but also using the Orbitrap technology [66]. Even more, Rao et al. [67] showed that this technology can be implemented on the conventional platform of hybrid instrument LTQ-Orbitrap. MSI technology would be useful in diagnosing difficult melanocytic lesions in skin cancer [68]. Lazova and Seeley showed that this technology could differentiate Spitz nevi (an uncommon form of skin mole) from Spitzoid malignant melanoma. Even more, in a recent study, the same author reported a histopathological misdiagnosed, congenital malignant melanoma that on a more detailed look by MSI revealed evidence of a congenital melanocytic nevus with proliferative nodules [69]. These reports underline the potential impact that mass spectrometry could have in the clinic, as the next focus is on the clinical applications of this technology [70].

BRAFV600E/K mutation is a major oncogenic driver with a frequency of 40–60% between patients diagnosed with advanced melanoma cancer. This mutation is commonly associated with MAP kinase pathway activation. In this sense BRAF blockade therapy is usually combined with MEK inhibitors. In BRAF wild-type cells, BRAF inhibitors are likely to lead to MAP kinase activation and thus are not advised for association [70, 71].

NRAS mutant melanoma is known as the second most common mutation observed in melanoma. Recent studies indicated that stage IV NRAS patients have a shorter overall survival rate than those with wild-type form [72].

New melanoma-related biomarkers were discovered using proteomic analysis of five melanoma cell lines. Two of these are represented by vimentin and nestin, intermediate filament proteins involved in cell migration, apoptosis, and cytoskeletal organization. Vimentin may play an important role in migration and invasion. Studies indicated that those cells close to blood vessels present higher levels of vimentin. Together with vimentin, the nestin level can be correlated with the aggressiveness of the tumor. Both were validated as melanoma-related proteins using samples from 40 patients with different stages of melanoma. These two proteins can be viewed as potential biomarkers to distinguish between different forms of melanoma [73].

A recent study reveals a potentially new biomarker, SerpinE2, responsible for the invasive phenotype of melanoma slow-cycling cells. Proteome analysis reports two other proteins enriched in melanoma cells, PDGFRL and BMP1 [47].

Microphthalmia-associated transcription factor (MITF) expressed by melanoma cells is known as a driver of melanoma progression. It is involved in multiple biological processes like proliferation and differentiation. Due to its high sensitivity and specificity, this transcription factor is frequently used to differentiate highly proliferative melanoma cells from poorly invasive population [74].

Al-Ghoul et al. performed a comparative analysis of proteins expressed by either primary melanoma cells or metastatic cells. Their results indicate higher abundance of metabolic proteins upregulated in melanoma cells. About 131 proteins have a higher level than normal in these cells. Cyclophilin A, a protein linked to cancer progression and inflammatory diseases, was also identified to be upregulated in metastatic melanoma [75].

Histopathologic characteristics of 10 samples originating from patients with stage III metastatic melanoma were correlated with protein expression obtained using mass spectrometry. Authors define four proteins positively correlated to melanoma tumor, and among them are melanoma cell adhesion molecule (MUC18), melanoma chondroitin sulfate proteoglycan (CSPG4), melanoma-associated antigen D2, and melanocyte protein (melan A) [76].

S-100β serum level can reflect clinical stage. In general its concentration is less than 0.10 μg/L in healthy population. Usually S-100β level is used to monitor the response to treatment in patients with metastatic tumor. A second prognostic biomarker largely used is LDH, a well-known prognostic factor commonly used in those patients diagnosed with stage IV melanoma [46].

Based on the assumption that eIF4A1 and eIF4E are overexpressed in cancer cells, Joyce et al. tried to elucidate the possible mechanism related to invasive phenotype of cancer cells. Silencing either eIF4A1 or eIF4E reduces melanoma proliferation. Recent findings reveal that eIF4A1-depleted melanoma cells and eIF4E-depleted melanoma cells may present quite different proteomes, which may lead to hypothesis that they induce invasive phenotype through alteration of different molecular mechanisms. In order to perform this comparative proteome analysis, a MS operating in MS3 mode was used [77].

5. TOF MS applications to biomarkers of human glioblastoma

Discovery and implementation of these fragmentation techniques, together with ESI and MALDI, revolutionized the field of MS, bringing significant contributions to the development of its analytical potential and its successes toward the topic of great interest, such as neoplastic transformations.

Accounting for 60% of all primary brain tumors, astrocytomas are already the focus of research for a few years [78]. Of these, glioblastoma multiforme (GBM), classified by World Health Organization (WHO) as having the highest malignancy degree, grade 4, is one of the most common and aggressive malignant brain tumor [79]. Treatment of GBM consists of surgical resection, radiation, and chemotherapy. Unfortunately, patients with this type of tumor have a median survival of approximately 14–15 months from the diagnosis [80, 81], with an estimated 5-year survival rate of 5%.

This poor prognosis could be allied to the richly neovascularized solid tumor profile, with many upregulated angiogenic and mitogenic factors [82], in other words, the high infiltrative and diffuse spreading of the GBM cells over long distances in the brain.

In recent years considerable efforts were invested in understanding the molecular mechanisms governing GBM tumorigenesis and also for the discovery of novel and more efficient therapeutic targets [83].

Astrocytomas and especially GBM were intensively addressed by proteomic studies, particularly by two-dimensional polyacrylamide gel electrophoresis (2D-PAGE), followed by protein digestion in a bottom-up approach. Nevertheless, the high sensitivity and reproducibility of MS, the need for significantly fewer test materials, as well as the MS relative simple working procedure over 2D-PAGE evolved MS as a powerful tool in screening cellular and tissue protein expression profiles (**Figure 3**) [84].

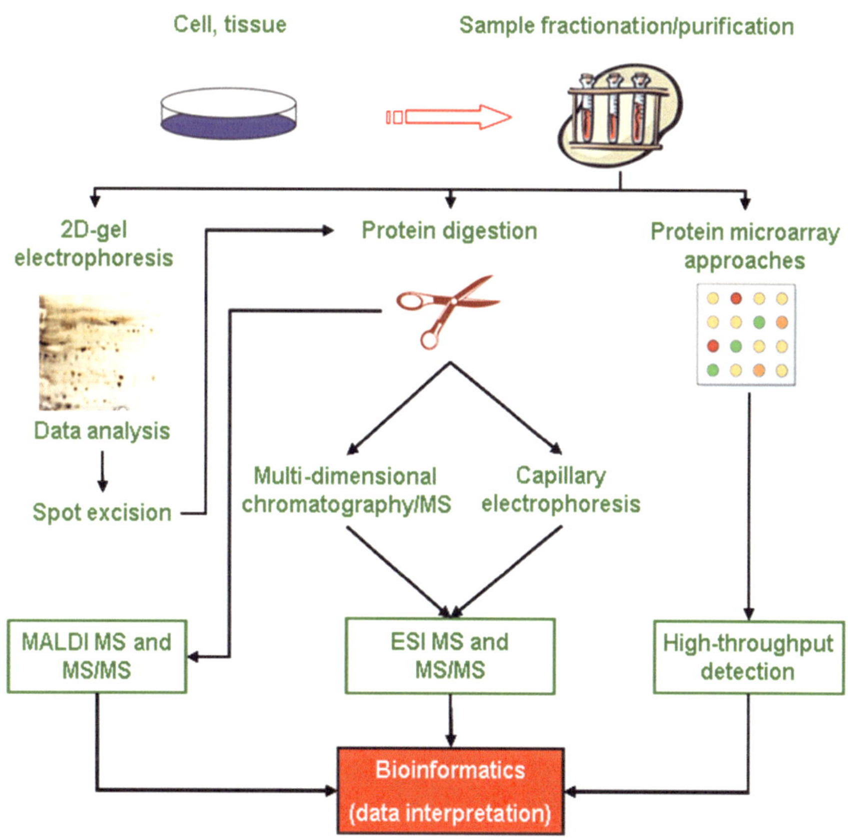

Figure 3.
Basic analytical strategies in proteomics.

In 2005, the proteome investigation in a single GBM tissue sample conducted by Wang et al. employing the combination of capillary isoelectric focusing (CIEF) and reversed-phase chromatography for multidimensional peptide separation hyphenated to QTOF MS led to the identification of a total of 6866 fully tryptic peptides, corresponding to 1820 distinct proteins [85].

MALDI-TOF-MS protein profiling has been used by Schwartz et al. to determine protein expression patterns that distinguish primary gliomas from normal brain tissue and one grade of gliomas from another, with high sensitivity and specificity. From the total of 162 tissue samples from 127 patients, including 19 patients undergoing resective surgery for nonneoplastic disease, 29 grade 2, 22 grade 3, and 57 grade 4 glioma patients that were analyzed, over 100 potential, tumor-specific biomarkers were identified. With their approach, they have distinguished glioma grades with high accuracy ranging from 76% to 97%, encountering some difficulty in distinction between the WHO grade II and III tumors [86].

A complex protein data set for GBM from direct tissue analysis including 2660 proteins with a significant number of peptide hits in total and 1401 validated protein hits with at least two unique peptide identifications was provided by the group of Huber CG [87]. Their novel methodology was based on the application of ion-pair RP chromatography to fractionate intact proteins from the tumor tissue prior to identification by MALDI-TOF/TOF-MS^2, as well as the combination of intact protein fractionation with shotgun proteomics to discover the most complex data set of proteins.

Considering the limited accessibility to healthy brain tissue samples, especially from the same patient, achieving a normal vs. diseased brain comparison at the proteome level is challenging. Most of the time, for such investigations, control brain tissues are obtained by surgery from patients with epilepsy or autopsy. In order to identify and characterize proteins that are differentially expressed in GBM and gather information on interactions and functions that lead to this condition, ten GBM tumors (patients from 48 to 67 years) and ten epileptic brain tissue (patients from 21 to 61 years) samples were analyzed by Mirza and Shamim [88]. By SDS PAGE fractionation with internal DNA markers followed by liquid chromatography-tandem MS [87] on a LTQ-Orbitrap Velos, differential protein expression in GBM was identified. These proteins, highly specific to GBM irrespective of their location in the brain (right or left temporal region of the brain), were further characterized by Ingenuity Pathway Analysis (IPA) to assess protein interactions, functions, and upstream regulators. Such an approach provided several other upregulated proteins, such as SERPH, PDIA1, CERU, TENA, VTNC, APOE, LEG1, HRG, and FKBP5, identified to be involved in tumor progression, aggressiveness, and invasion in GBM. Besides the known GBM-associated proteins, CLIC4, NP1L1, IGKC, TAGL2, and YES, previously correlated with processes promoting cancer progression, were identified here as novel potential biomarkers of GBM [83].

Despite the significant efforts invested over the past decades in finding an appropriate treatment for GBM, almost all GBMs are ultimately resistant to existing treatment and inevitably recur [89]. Though a multimodal approach including surgery, postoperative radiotherapy, and chemotherapy is applied, the prognosis for patients with GBM remains rather poor. Hence, there is an urgent need for a better understanding of the molecular mechanisms associated with GBM, improvement of conventional treatment modalities, and/or development of new and more efficient ones. A first step in breaking the evolution and/or recurrence of this fast-growing tumor is the development and application of new, rapid, and accurate approaches to assess the treatment of biological effects, given that currently, the treatment results are radiologically evaluated several months after treatment.

Surface-enhanced laser desorption/ionization (SELDI) TOF MS demonstrated its ability in detecting the changes in protein patterns in GBM tumor tissue following radiotherapy. In the study reported by Wibom et al. [90], rat BT4C glioma cells were implanted into the brain of two groups of 12 BDIX-rats, one of the groups receiving radiotherapy. The expression of proteins in normal and tumor brain tissue, pooled at four different time points after irradiation, was investigated using SELDI TOF MS by principal component analysis (PCA) and partial least squares (PLS). The variations over time in protein expression identified by the authors, together with the tumor progression in vivo and, last but not least, the rapid and significant changes in the protein expression following irradiation, allowed the irradiated tumors to be clearly distinguished from the nontreated tumors. Therefore, the authors consider that the 77 identified signals characterized by different intensity levels in irradiated vs. nonirradiated tumors validate SELDI TOF MS as a method of choice in elucidating biological events induced by radiation which can be used as markers for monitoring the efficacy of radiation treatment in malignant glioma [90].

The pioneer study in which MRI was used to guide proteomic analysis with SELDI TOF MS in human GBM in order to demonstrate a correlation between imaging patterns and protein expression profiles belonged to Hobbs et al. [84]. Contrast-enhanced (CE) magnetic resonance imaging (MRI) represents an important diagnostic tool for correlating molecular pathophysiology with the clinical management of disease; it can supply elevated spatial resolution and molecular signatures of both healthy and diseased tissues in solid tumors. For the identification of CE and nonenhanced (NE) regions of the tumor in the four patients previously diagnosed with GBM, intraoperative stereotaxis during surgical resection in conjunction with MRI was used. Further, Hobbs et al. investigated the profile of over 100 proteins and peptide species within GBM tumors and compared the protein profiles between the CE and NE regions across four patients with MR images and confirmed diagnosis of GBM. This CE and NE comparative approach on the GBM proteomic fingerprint revealed qualitative and semiquantitative proteomic pattern differences, hinting an impaired gene expression profile that correlates with detectable tissue imaging parameters [84].

Although noninvasive imaging techniques have improved the neuroradiological diagnostic accuracy, in some cases, comprehensive specificity for discrimination of brain tumors and detection of minor differences in tumor size and behavior are difficult using imaging approaches [91]. Besides, tumor tissue extraction through biopsy or resection is not always feasible. For such reasons, identification of reliable biomarkers in the blood would facilitate the management process of GBM patients, allowing an early diagnosis, establishment of surgical intervention plans, and monitoring of the disease course and the treatment response.

Quantitative comparisons of the plasma proteomes of 14 GBM patients and 15 healthy controls using sequential window acquisition of all theoretical fragment ion spectra (SWATH) MS analyses were conducted for identifying potential biomarker candidates in plasma of GBM patients. SWATH MS, characterized by high reproducibility and reliability of quantitative information, combines a highly specific data-independent acquisition [38] method with a novel targeted data extraction strategy to mine the resulting fragment ion data sets [92].

Being a label-free analysis, SWATH MS was employed by Miyauchi et al. in conjunction with LC-MS/MS using a Triple TOF 5600 for generating a protein quantification method, called quantitative targeted absolute proteomics (QTAP) [91]. With the current approach, the authors not just estimated the origin of upregulated biomarker candidates in GBM plasma but also have examined if the upregulated biomarker candidates were elevated in GBM tissues, and if the biomarkers were detectable in cyst fluid. Additionally, the relationships between the concentrations of biomarker candidates in plasma and the clinical presentation (tumor size, progression-free survival time (PFS), or overall survival time (OS)) of GBM patients were inspected for establishing the relationships between biomarker candidates and GBM biology. While LRG1, C9, CRP, SERPINA3, and APOB were identified by SWATH MS and QTAP analyses as upregulated biomarker candidates, compared with the healthy plasma, GSN, IGHA1, and APOA4 were identified as downregulated biomarker candidates (Miyauchi et al.). All the candidate proteins, except for CRP, were found to be highly expressed in cytosol of GBM tissues vs. noncancerous brain tissues, while GSN was identified as decreasing also in cerebrospinal fluid (CSF). Although valuable information was gained by SWATH MS and QTAP, in order to determine the feasibility of their findings in clinical application, the authors recommend further investigations using plasma from patients with postoperative GBM, glioma, and other various types of cancer [91].

Since in 2010, there was no CSF biomarker in clinical use in malignant brain tumors, neither for diagnostic nor for other purposes. The detection of recurrence or reaction to adjuvant therapy and the exploration of CSF originating from GBM patients in respect to peptide profiling were not experimented as well. Eleven CSF samples collected by lumbar puncture from four female and seven male patients with a single supratentorial GBM manifestation (median 68 years, range 43–79 years) and 13 samples from four female and nine male participants with mostly spinal canal stenosis, and used as control, were involved in a study conducted by Schuhmann and colleague [93]. The peptides, previously separated by RP HPLC in 96 fractions per sample, were resuspended in matrix solution consisting of α-cyano-4-hydroxycinnamic acid and L-fucose (co-matrix) in 0.1% acetonitrile/trifluoroacetic acid (1:1 v/v) and screened by nanoESI-QTOF-MS/MS. To detect possible marker candidates, the two peptidomes were compared using the differential peptide display approach. Additionally, patient data including age, duration of symptoms, and the morphological MRI data were correlated to the mass spectrometric signal intensity of the peptides in CSF using Spearman's rank correlation coefficient. More than 6000 peptide signals, corresponding to at least 2000 different peptides, were remarked in the peptide master displays of the two peptidomes. Of these, four CSF peptides, specific *C*-terminal fragments of α-1-antichymotrypsin, osteopontin, and transthyretin as well as a *N*-terminal residue of albumin, significantly distinguished GBM from controls in all applied statistic tests. Although these molecules are components of normal CSF, none of these peptides or their precursors was previously reported as significantly elevated in CSF of GBM patients.

Although neuronavigational guidance by MRI or positron emission tomography (PET) aids in localizing the tumor tissue, the brain shift and tissue deformation may negatively influence the accuracy of resection [94, 95]. An option to overcome these limitations is the application of intraoperative imaging; however, besides being expensive they increase the surgery time [95].

Since the introduction of fluorescence-guided surgery (FGS) in 1998 by Stummer et al. subsequent to proagent 5-aminolevulinic acid (5-ALA) treatment, contrast enhancement by a fluorescent dye has progressed and represents today an important and powerful technique in GBM resection [94, 95]. After oral application of 5-ALA approximately 3 h before anesthesia, in a dose of 20 mg/kg body weight, the small molecule crosses the abnormal blood-brain barrier into the peritumoral tissue and accumulates within the tumor. As a precursor of heme B, in an intermediate step of the heme biosynthesis cascade, 5-ALA is converted to protoporphyrin IX, a red fluorescent at 635 nm upon blue light excitation. Since high-grade gliomas are known to be characterized by augmented cell proliferation, tumor cell density and microvessel density, the greater fluorescence intensities generated by accumulation of high cellular PpIX concentrations, improve differentiation of tumor from healthy brain tissue during intraoperative fluorescence microscopy [95–97]. Nevertheless, in some cases, false-positive and false-negative fluorescence results cannot be excluded. Hence, molecular and elemental MSI was introduced as a complementary technique to examine the potential chances and boundaries of fluorescence diagnosis. Within the molecular MSI domain, MALDI MS was demonstrated to have a high applicability, even of FGS for meningioma, since it accomplishes the investigation of intact molecules by soft ionization [98].

A comprehensive study on the fluorescent drug PpIX and the biosynthesis end product heme B, based on MALDI MSI followed by quantitative Fe analysis via laser ablation inductively coupled plasma (LA-ICP) MS, was performed by Kröger et al. in the group of Karst [95]. Three GBMs after and two GBMs without 5-ALA administration, one gliosarcoma (GSM), one low-grade glioma, and two reactive brains were characterized using an ion trap-TOF mass analyzer and α-cyano-4-hydroxycinnamic acid as matrix for MALDI.

With their approach, Kröger et al. enabled the detection of PpIX distribution in GBS, where high tumor density exhibited intense PpIX signal during surgery. Heme B and enhanced Fe accumulation was detected only in regions of blood vessels and hemorrhage, corroborating the hampered transformation from PpIX to heme B in GBM tissue. In the case of GSM, the absence of PpIX accumulation determined by the nonfluorescent appears during FGS, generated a false-negative fluorescence diagnosis [95].

Although initially reported in 1895, GSM, a rare primary neoplasm of the CNS, did not gain a broad acceptance until 1955, when three cases of patients with this malignancy were described in detail [99]. Thus, hitherto, only a limited number of case reports and researches were published, and from those, just a few were targeting the identification of potential biomarker species in GSM.

Defined by the WHO as a variant of GBM, accounting for about 2% of all the GBM, GSM affects preponderantly the adult population in the fourth to the sixth decade of life, more frequently males than females (male/female, 1.8:1) [79]. Consisting of alternating areas displaying glial and mesenchymal (sarcomatous) differentiation, GSMs are typically located in the cerebral cortex, involving the temporal, frontal, parietal, and occipital lobe in decreasing frequency [100]. The epidemiology and natural history of this rare malignancy appears similar to GBM, with a slightly greater predisposition for temporal lobe involvement. As for GBS, the prognosis for GSM after surgical resection and chemotherapy is limited, even inferior to those observed in GBM patients, with a median survival of 11–12 months, with less than 10% survival after 2 years following diagnosis [101]. Such a reduced life expectancy may be caused by the widespread penetration of the GSM cells into the brain tissue that makes the tumor cells slightly inaccessible to treatment methods and so the treatment fails [102].

It is broadly accepted that abnormalities in glycoconjugate glycosylation pathways are responsible for atypical cell surface—glycocalyx—molecular structure, and the latter one accompanies malignant alteration of cells. Since the cell surface carbohydrates are involved in a variety of interactions of the cell with its extracellular environment, any change in their expression cause tumor cell migration, metastasis, and invasiveness [103].

In view of the range of glycosyl epitopes that designate tumor-associated antigens [104, 105], some of them endorsing proliferation and metastases and some suppressing the tumor progression [106], surveying the changes in the profile of the glycoconjugate molecules liable for such effect, represents an alternative to the proteomic studies. Atypical glycosylation associated with oncogenic transformation was first confirmed with gangliosides [107]. Gangliosides (GGs) are sialic acid-containing glycosphingolipids integrated into the outer leaflet of the cellular membrane bilayer and predominantly enriched in microdomains.

Within the intermolecular interactions and through glycosynapses, GGs are known to regulate the cell-cell recognition and intercellular adhesion and to modulate signal transduction pathways. The stimulation of cell growth inhibition and cell differentiation and/or apoptosis of certain GGs is well demonstrated for nearly two decades [108]. For instance, the cell number in primary cultures of high-grade human GBM, ependymomas, mixed gliomas, astrocytomas, oligodendrogliomas, and gangliogliomas, as well as of the rat 9 L cell GSM cell line, was shown to decrease following treatment with GM3. In the same time, it had a slight effect on the cell number in cultures of a normal human brain [100]. Given the correlation of different malignancy grade and median survival time with GG composition and content in different types of glioma tumors, GGs can be considered as tumor markers and used as therapeutic targets or vaccine development [109].

A diminished amount of complex GGs in tumors is connected to the dedifferentiation processes, reflecting the GG pattern alteration in ontogenesis during differentiation of nervous tissue: an increased content of complex GGs and a reduced volume of simple GGs [110, 111].

Together with the immunochemical and immunohistochemical methods, several other biophysical techniques were introduced over the years for mapping and structural analysis of GGs in different gliomas, such as HPLC, infrared spectroscopy, and confocal microscopy via anti-GG monoclonal antibodies [112–114]. Additionally, the implementation of miniaturized, integrated devices for sample infusion into MS simplified the laborious chemical and biochemical MS strategies by reducing sample handling and sample loss, eliminating potential cross-contamination, and increasing ionization efficiency and quality of spectra as well as the possibility to perform unattended, high-throughput experiments [115]. The world's first fully automated nanoelectrospray system and perhaps the most popular and widely used device is the NanoMate robot produced by Advion BioSciences (**Figure 4**), in which 100 or 400 nanospray emitters are integrated onto a single silicon substrate, from where electrospray is established perpendicular to the substrate. Considering the (i) automated infusion into MS at low flow rates in the nanoliter range (only 50–100 nL/min), (ii) minute amount of sample, (iii) technical quality of the nanosprayers, (iv) reproducibility of the experiment, (v) formation of preponderantly multiply charged ions, and (vi) minimized in-source fragmentation of labile groups attached to the main structural backbone, in some cases, chip-based devices not just eliminated the need for separation prior to MS analysis but also provided more robust and quantitative analyses than LC- or CE-MS [116].

The first investigation of GG pattern in a human GSM specimen by sensitive and accurate MS methods, complemented by HPTLC analysis was conducted in 2006 [100]. The current approach was developed and optimized in order to detect and structurally characterize the tumor-associated species, which might serve as potential diagnostic markers or specific target molecules for the production of antitumor therapeutic agents. Primarily, the densitometric analysis conducted in two complex GG mixtures, one originating from a 47-year-old male presenting GSM tumor and one from adult frontal cortex used as control, indicated a severalfold decrease (7.4×) of the total GG content in GSM vs. normal human brain (**Figure 5**), which was in agreement with previous reports on other glioma types. Further, by employing chip-based nanoESI QTOF MS and ultrahigh-resolution nanoESI FTICR MS, Vukelić et al. demonstrated for the first time the reduction in the total GG content and the altered pattern in GSM vs. control tissue [100]. While GM1a, GD1a, and GD1b abundances were very low in GSM, GD3 species were found to account

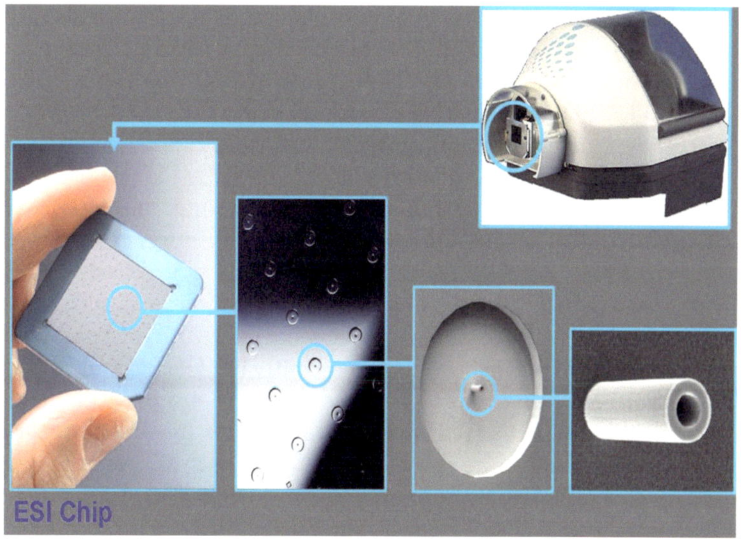

Figure 4.
NanoMate robot with fully automated chip-based ESI (courtesy of Advion BioSciences).

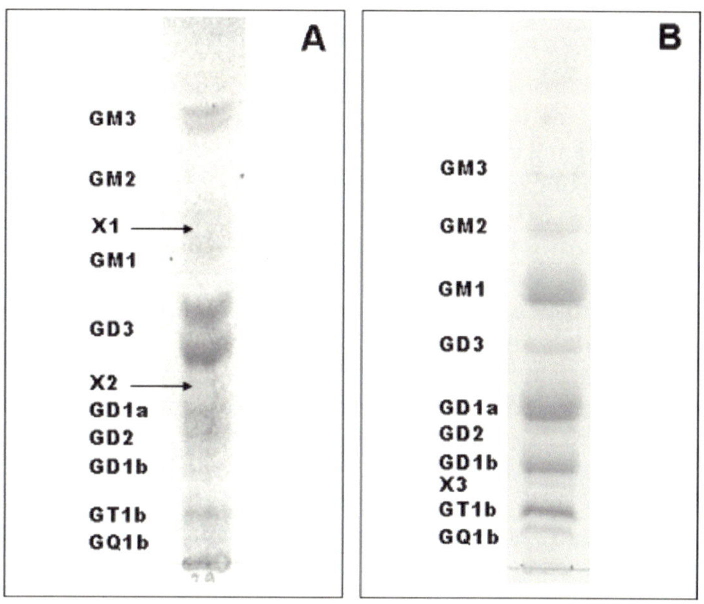

Figure 5.
Compositions of GGs isolated from human gliosarcoma specimen (A) and normal human frontal cortex (B), as revealed by HPTLC followed by densitometric scanning. Reprint with permission from Vukelić et al. [100].

for nearly 50% of the total GG content, being the most emphasized fraction and the only one expressed in higher absolute concentration than in normal cortex (approximately 128%). Therefore, the authors hypothesized that the combination of both lower overall biosynthetic rate, due to change in expression of certain glycosyltransferases, and higher turnover rate is responsible for such a pattern. At the same time, the considerably higher amount of potentially proapoptotic GD3 than of the *O*-acetyl GD3 species observed by the authors supported the assumption of Kniep et al. [117], according to which, *O*-Ac-GD3 could by itself be responsible for the protection of tumor cells from apoptosis. Moreover, the high sensitivity and mass accuracy of QTOF and FTICR MS and MS/MS permitted them the assignment of unusual GG minor species. Among these species, GM4, Hex-HexNAc-nLM1, Gal-GD1, Fuc-GT1, GalNAc-GT1, *O*-Ac-GM3, di- *O*-Ac-GD3, *O*-Ac-GD3, and *O*-Ac-GT3, considered as brain developmental antigens [118], were not reported previously as glioma-associated structures [100]. The fully CID sequencing at low energies of *O*-Ac-GD3(d18:1/20:0), exhibiting higher abundance in GSM vs. normal human brain, provided structural evidence to postulate a novel *O*-Ac-GD3 isomer, *O*-acetylated at the inner Neu5Ac-residue, previously not structurally confirmed.

In order to reveal the role of GG as tumor-associated antigens, a detailed depiction of native GG mixtures from normal human brain (NB), GBM and corresponding peritumoral tissue [119], was conducted recently by HPTLC and MS [110]. Particular attention was paid to the GG expression in PT, since the biochemical processes occurring within this area are believed to be responsible for aggressive infiltration of tumor cells into the surrounding brain tissue and therefore for high incidence of GBM recidivism and poor prognosis of GBM [110, 120].

While the frontal cortex of a 54-year-old who died in a traffic accident was used as control, the GBM and the corresponding PT were collected from a 70-year-old female patient, whose brain MRI showed an infiltrative tumor lesion measuring 47 × 43 × 40 mm in the right frontal and insular region. This approach revealed distinctive changes in GG expression in GBS. Total GG content in GBS was approximately five times lower than in the NB, while approximately two times lower than in the PT. For MS analyses, the extracted GG mixture dissolved in methanol up to 0.8 mM concentration was infused into a Bruker amaZon ETD ion trap system. Whereas predominantly C18 chains were found to characterize the NB and PT, shorter (C16) and longer (C22, C24), unsaturated (C24:1) and with odd number carbon atom (C17, C19) fatty acids residues were detected in GBS [110]. The most abundant GG in GBS (accounting for 53% of the total GG content) was GD3(d18:1/18:0), followed by GD3(d18:1/24:0) that was exclusively detected in GBS tissue. Likewise, proportions of GM3, GM2, GD2, and O-Ac-GD3/nLM1 were much higher in GBS than in the PT and NB tissue. In contrast, GM1, GD1a, GD1b, and GT1b constituted the most abundant GG structures in NB, their amount being considerably lower in GBS compared to NB and PT. The content of GGs modified by *O*-acetylation, most often of the sialic acid, was also found altered in GBS. The previous reports, according to which *O*-acetylation exhibits antagonistic effect on GD3 function, thus modulating its activities [121] and protecting the cell from its proapoptotic activity [122], herein confirmed (i) the high intensity of *O*-Ac-GGs in healthy brain tissue, (ii) the low concentrations in PT, and (iii) the detection of only one ion corresponding to *O*-Ac-GT1(d18:1/18:0) in GBM sample. *O*-Ac-GD1 was detected in NB and PT, but not in GBS, while *O*-Ac-GD3 species were found exclusively in GBS.

6. Applications of MS in pancreatic cancer (PDAC)

Although it has a lower incidence and prevalence compared to other cancers, PDAC is one of the deadliest, patients affected by PDAC having a very low life expectancy. This is due to the low success in the detection of this disease, as well as to a high resistance to therapies. Therefore, it is of major interest to set up biomarkers or biomarker panels dedicated to the earlier detection of the disease, as well as biomarker panels for the prediction of therapy outcomes. Many works have been dedicated to these aspects, including approaches based on genomics, proteomics, metabolomics, etc., or based on combination of such techniques.

Liu et al. used a combination of MS-intensive methods such as isobaric tags for relative and absolute quantitation with two-dimensional liquid chromatography-tandem mass spectrometry (iTRAQ-2DLC-MS/MS) and 1D-targeted LC-MS/MS, on serum samples from healthy people (normal control, NC), patients with benign diseases (BD), and PC patients to identify novel biomarkers of PC [123]. From more than 1000 identified proteins, 142 were identified as differentially expressed. For the diagnosis of PC, a novel biomarker panel consisting of apolipoprotein E (APOE), inter-alpha-trypsin inhibitor heavy chain H3 (ITIH3), apolipoprotein A-I (APOA1), and apolipoprotein L1 (APOL1), in combination with CA19-9, statistically improved the sensitivity (95%) and specificity (94.1%), outperforming CA19-9.

Radon et al. used the LC-MS/MS analysis on urine samples from patients and controls and identified LYVE-1, REG1A, and TFF1 as candidate biomarkers [119]. Further the selected biomarkers were investigated by ELISA over a larger cohort (192 PDAC patients and 87 normal urine samples) and provided an AUC for the panel of 0.90, with the detection of PDAC in stages I–II.

Nigjeh et al. [124] applied a methodology of data-independent acquisition (DIA) [38] that included sample preparation with high reproducibility and separation by LC-MS/MS; they estimated the applicability in the detection of pancreatic cancer, based on a plasma spectral library consisting of over 14,000 identified peptides obtained from more than 2300 plasma proteins. The reliability of quantification was examined for the identified peptides, in constant retention times and signal intensity. The linear dynamic range and the lower limit of quantification were evaluated and pointed toward a critical role of the sample complexity for the optimization of DIA settings. Validation of the assay based on cohort of clinical plasma demonstrated the robustness and a unique advantage for targeted analysis of plasma samples for biomarker development.

In the study of Yoneyama et al. [125], LC-MS/MS proteomics was applied for the quantitation of plasma proteins, in combination with specific antibodies. The results showed that insulin-like growth factor-binding proteins (IGFBP)2 and IGFBP3 display the potential to distinguish early-stage PDAC patients from normal controls and that IGFBP2 was increased in intraductal papillary mucinous neoplasms (IPMNs). Moreover, PDAC diagnosis based on a combination of IGFBP2 and IGFBP3 and carbohydrate antigen 19-9 (CA19-9) is significantly more efficient compared to only CA19-9. These prove that IGFBP2 and IGFBP3 can represent supplemental or compensatory biomarkers to CA19-9 and by using this biomarker combination may improve the prognosis of PDAC patients.

Castillo et al. [126] examined analyzed exosomal surface proteins, using a combination of liquid chromatography and mass spectrometry. Simultaneously, PCR was applied on samples from 74 patients (a total of 136 exosome samples) to detect baseline KRAS mutation rates in patients subjected to therapy. The exosome proteomic analysis identified a set of biomarker candidates for PDAC: CLDN4, EPCAM, CD151, LGALS3BP, HIST2H2BE, and HIST2H2BF. In 44.1% of the patients under active therapy, KRAS mutations were detected in total (unsorted) exosomes; the yield increased to 73.0% when exosome capture was conducted based on selected biomarkers.

7. Applications of MS in colorectal cancer

Kirana et al. used a combination of 2D-DIGE and MALDI-TOF mass spectrometry for 125 tumor tissue samples in different TMA stages of CRC. Out of 55 proteins with differential expression, a group of ten protein biomarkers was selected, comprising HLAB, protein 14-3-3β, LTBP3, ADAMTS2, JAG2, and NME2; they were significantly associated with clinical parameters relevant for tumor progression, invasion, and metastasis. Kaplan-Meier survival curve demonstrated the marked expression of six proteins that could be associated with improved survival from CRC. The level of expression of HLAB, ADAMTS2, LTBP3, JAG2, and NME2 on tumor cells was associated with progression and invasion, metastasis, and CRC-specific survival; all these may be useful biomarkers for stratification of CRC patients regarding low and high risk of metastasis. A combination of microdissection with 2D-DIGE and MALDI-TOF MS was demonstrated to be useful in the identification of biomarkers for the prediction of the risk of liver metastases [127].

Suwakulsiri et al. applied quantitative mass spectrometry to examine the proteomes of two classes of extracellular vesicles—exosomes and shed microvesicles (sMVs) [133]. Two populations of sMVs were investigated—released from primary (SW480) and metastatic (SW620) human isogenic-colorectal cell lines. One thousand two hundred ninety-five and one thousand three hundred proteins were identified in SW480-sMVs and SW620-sMVs, respectively, using quantitative mass spectrometry. Analysis of Gene Ontology identified processes of "cell adhesion" (CDH1, OCLN, CTN families), "signaling pathway" (KRAS, NRAS, MAPK1, MAP2K1), and "translation/RNA related" (EIF, RPL, HNRNP families) in both types of sMVs. Remarkably, SW480- and SW620-sMVs displayed distinct signatures; SW480-sMVs was enriched in ITGA/B, ANXA1, CLDN7, CD44, and EGFR/NOTCH signaling networks, while SW620-sMVs are higher in PRKCA, MACC1, FGFR4, and MTOR/MARCKS signaling networks. Jimenez et al. evaluated the proteins in small and large extracellular vesicles (SEVs and LEVs, respectively) [128], applying isobaric tag for relative and absolute quantitation liquid chromatography (iTRAQ-LC) [87] tandem mass spectrometry (MS) on EVs from colon cancer cell line (SW-480) and the lymph node metastatic line (SW-620) and a CRC patient primary culture. Bioinformatic analyses showed that SEVs contain higher levels of proteins involved in cell-cell junctions, cell-matrix adhesion, exosome biogenesis, and diverse signaling pathways and that LEVs contain higher levels of proteins associated with the biogenesis of ribosomes and RNA biogenesis and processing and with metabolism.

Kit et al. [129] analyzed the patterns of expression for proteins implicated in cell signaling, using paired samples tumor/nontumor colon cancer patients, with the purpose of finding protein clusters capable of differentiation between patients having non-metastatic and metastatic colon cancer. They found in tumoral mucosa nine proteins upregulated significantly: among these are protein kinase C gamma, c-Myc, MDM2, and pan-cytokeratin, while GAP1 was significantly downregulated. Pan-cytokeratin and APP appeared upregulated in tumor compared to nontumor tissue and were thus included in the predictive cluster for discrimination of cancer type. All applied methods of regression/clustering confirmed the presence of increased concentrations of S-100b and phospho-Tau-pSer199/202, and their predictor value for non-metastatic colon cancer. Further investigations are required for the validation of potential protein markers for colon cancer development and metastatic progression.

Lee et al. [130] used liquid chromatography-tandem mass spectrometry (LC-MS/MS) analysis for identification of potential biomarkers in colon cancer. They identified tetraspanin 1 (TSPAN 1) as a potential noninvasive biomarker.

Liu et al. [123] applied MADLI-TOF-MS for the analysis of the N-glycome of IgG in patients with colorectal benign tumors, colorectal cancer, and normal individuals. The results identified nine IgG N-glycans that were differentially expressed in patient groups. Moreover, five of them were significantly modified in CRC patients in all TNM stages. Principal component analysis (PCA) indicated evident differentiation between benign, cancer patients and normal individuals. The ROC analysis suggests that these five IgG N-glycans were correlated with the progression of

CRC. Analysis of IgG N-glycosylation suggests that core-fucosylation, sialylation, and sialo core-fucosylation were probably correlated with development of CRC.

Hu et al. [131] used stable isotope labeling by amino acids in cell culture (SILAC) technology. Thus, HCT116-I8 and HCT 116 cells were cultivated with either "light media" (Arg0, Lys0) or heavy media (Arg10, Lys8), for 7 days, followed by lysis of cultures, protein digestion, and MS analysis on an LTQ-Orbitrap mass spectrometer. The outcome of the study was Cdc42-Cdc42BPA signaling as prognostic biomarker and therapeutic target for colon cancer invasion. Also, the authors discovered a decreased expression of E-cadherin and increased expression of vimentin in highly invasive cell lines.

Alvarez-Chaver et al. [132] investigated the proteomic profile of tumor and mucosa in CRC, using a combination of 2D electrophoresis and mass spectrometry, revealing nucleotide-diphosphate kinase as a candidate biomarker. They finally used an ELISA kit for the validation of the biomarker.

8. Conclusions

Mass spectrometry provides powerful instruments for the analysis of biological systems, with dedicated instruments for different classes of analytes, such as proteomes, metabolomes, genomes, transcriptomes, etc. The power of such systems is usually backed by powerful bioinformatic tools, eventually playing at different levels (the "built-in" instrument that controls effectively the instrument), and the post-analytical bioinformatics tools that help in in-depth analysis and interpretation of findings.

We have tried to illustrate the state of the art concerning the available technology in MS for proteomics, and we have illustrated these with application in biomarker discovery in melanoma, glioblastoma, pancreatic cancer, and colorectal cancer.

Similar approaches are developed also for other solid cancers—such as gastric, hepatic, pulmonary, ovarian, or prostate—but, unfortunately, there is not enough space and time to approach them in this very chapter.

The core of the proteomic approaches is still based on two sample types: tissue (tumoral, peritumoral, and nontumoral—generally collected surgically or biopsies collected presurgically) and blood (most often, serum/plasma, more recently circulating tumor cells (CTCs) or circulating tumor stem cells (CTCS), and, in the last years, exosomes). Exosomes may represent an interesting option, since they bear significant cytosolic contents of the original cells (like nucleic acids and proteins) and also have a specific molecular surface signature of the originating cells. The major question on ECVs is represented by the amounts of such entities that can be recovered from small volumes of blood. Several other sample types were considered in the recent literature as sources for discovery of biomarkers and further as biological samples for diagnostics. One first example is represented by tears; it is considered that they have the overall load of plasma proteins, but devoid of the major ones (such as serum albumin), and could serve as biological material for diagnostics, prognostics, and patient stratification. Other such samples are saliva and urine; as in the case of tears, their collection is completely noninvasive.

Once one or several biomarkers are identified, next stages of pre-validation and validation are required, usually these will be achieved using non-MS techniques, such as ELISA, Western blot, or the use of multiplex platforms (Luminex, Meso Scale, etc.). One major problem in the progress toward clinical application is represented by the existence of equipment and detection kits validated for such applications, and this might represent a major problem for the future.

Finally, we want to encourage all who can implicate in cancer proteomics, thus accelerating the progress in knowledge.

Acknowledgements

The work for this publication was supported by projects PN-III-P4-ID-PCE-2016-0073, PN-III-P1-1.2-PCCDI-2017-0046, PN-III-P1-1.1-PD-2016-0256, PN-III-P1-1.1-PD-2016-1528, PN-III-P4-ID-PCE2016-0650, P_37_794-POC-A.1.14-E-2015, and 34 N/2019 - PN 19-41 05 01.

Author details

Radu Albulescu[1,7]*, Andrei Jose Petrescu[2], Mirela Sarbu[3], Alice Grigore[1], Raluca Ica[3], Cristian V. A. Munteanu[2], Adrian Albulescu[1,4], Ioana V. Militaru[5], Alina-Diana Zamfir[3], Stefana Petrescu[5] and Cristiana Tanase[6,7]

1 National Institute for Chemical Pharmaceutical R&D, Bucharest, Romania

2 Department of Bioinformatics and Structural Biochemistry, Institute of Biochemistry of the Romanian Academy, Bucharest, Romania

3 National Institute for Research and Development in Electrochemistry and Condensed Matter, Timisoara, Romania

4 "Stefan S. Nicolau" Institute of Virology of the Romanian Academy, Bucharest, Romania

5 Department of Molecular Cell Biology, Institute of Biochemistry of the Romanian Academy, Bucharest, Romania

6 "Victor Babes" National Institute of Pathology, Bucharest, Romania

7 Titu Maiorescu University, "Nicolae Cajal" Institute of Medical Research, Bucharest, Romania

*Address all correspondence to: radu_a1@yahoo.com

References

[1] Chen X, Wei S, Ji Y, Guo X, Yang F. Quantitative proteomics using SILAC: Principles, applications, and developments. Proteomics. 2015;**15**(18):3175-3192

[2] Zhu J, Warner E, Parikh ND, Lubman DM. Glycoproteomic markers of hepatocellular carcinoma-mass spectrometry based approaches. Mass Spectrometry Reviews. 2019. DOI: 10.1002/mas.21583

[3] Yamakawa Y, Kusuhara M, Terashima M, Kinugasa Y, Sugino T, Abe M, et al. CD44 variant 9 expression as a predictor for gastric cancer recurrence: Immunohistochemical and metabolomic analysis of surgically resected tissues. Biomedical Research. 2017;**38**(1):41-52

[4] Itoi T, Sugimoto M, Umeda J, Sofuni A, Tsuchiya T, Tsuji S, et al. Serum metabolomic profiles for human pancreatic cancer discrimination. International Journal of Molecular Sciences. 2017;**18**(4):pii: E767

[5] Kall L, Vitek O. Computational mass spectrometry-based proteomics. PLoS Computational Biology. 2011;**7**(12):e1002277

[6] Riley NM, Hebert AS, Coon JJ. Proteomics moves into the fast lane. Cell Systems. 2016;**2**(3):142-143

[7] Olsen JV, Schwartz JC, Griep-Raming J, Nielsen ML, Damoc E, Denisov E, et al. A dual pressure linear ion trap orbitrap instrument with very high sequencing speed. Molecular & Cellular Proteomics. 2009;**8**(12):2759-2769

[8] Rose RJ, Damoc E, Denisov E, Makarov A, Heck AJ. High-sensitivity orbitrap mass analysis of intact macromolecular assemblies. Nature Methods. 2012;**9**(11):1084-1086

[9] Snijder J, van de Waterbeemd M, Damoc E, Denisov E, Grinfeld D, Bennett A, et al. Defining the stoichiometry and cargo load of viral and bacterial nanoparticles by orbitrap mass spectrometry. Journal of the American Chemical Society. 2014;**136**(20):7295-7299

[10] Aebersold R, Mann M. Mass spectrometry-based proteomics. Nature. 2003;**422**(6928):198-207

[11] Gillet LC, Leitner A, Aebersold R. Mass spectrometry applied to bottom-up proteomics: Entering the high-throughput era for hypothesis testing. Annual Review of Analytical Chemistry. 2016;**9**(1):449-472

[12] Zhang Y, Fonslow BR, Shan B, Baek MC, Yates JR III. Protein analysis by shotgun/bottom-up proteomics. Chemical Reviews. 2013;**113**(4):2343-2394

[13] Tore V. Performing quantitative determination of low- abundant proteins by targeted mass spectrometry liquid chromatography. In: Mass Spectrometry. London, UK: IntechOpen; 2017. pp. 79-94

[14] Schwartz JC, Senko MW, Syka JEP. A two-dimensional quadrupole ion trap mass spectrometer. Journal of the American Society for Mass Spectrometry. 2002;**13**(6):659-669

[15] Boja E, Rivers R, Kinsinger C, Mesri M, Hiltke T, Rahbar A, et al. Restructuring proteomics through verification. Biomarkers in Medicine. 2010;**4**(6):799-803

[16] Jelonek K, Pietrowska M, Ros M, Zagdanski A, Suchwalko A, Polanska J, et al. Radiation-induced changes in serum lipidome of head and neck cancer patients. International Journal of Molecular Sciences. 2014;**15**(4):6609-6624

[17] Wilm M. Principles of electrospray ionization. Molecular & Cellular Proteomics. 2011;**10**(7):M111.009407

[18] Liuni P, Wilson DJ. Understanding and optimizing electrospray ionization techniques for proteomic analysis. Expert Review of Proteomics. 2011;**8**(2):197-209

[19] Sarbu M, Zamfir AD. Nanofluidics-based mass spectrometry. Applications for biomarker discovery in lysosomal storage diseases. In: Lungu M, Neculae A, Bunoiu M, Biris C, editors. Nanoparticles' Promises and Risks. Basel, Switzerland: Springer Nature AG; 2015. pp. 137-165

[20] Swiatly A, Horala A, Hajduk J, Matysiak J, Nowak-Markwitz E, Kokot ZJ. MALDI-TOF-MS analysis in discovery and identification of serum proteomic patterns of ovarian cancer. BMC Cancer. 2017;**17**(1):472

[21] Wada Y, Dell A, Haslam SM, Tissot B, Canis K, Azadi P, et al. Comparison of methods for profiling O-glycosylation: Human proteome organisation human disease glycomics/proteome initiative multi-institutional study of IgA1. Molecular & Cellular Proteomics. 2010;**9**(4):719-727

[22] Liu H, Zhang N, Wan D, Cui M, Liu Z, Liu S. Mass spectrometry-based analysis of glycoproteins and its clinical applications in cancer biomarker discovery. Clinical Proteomics. 2014;**11**(1):14

[23] Schulz S, Becker M, Groseclose MR, Schadt S, Hopf C. Advanced MALDI mass spectrometry imaging in pharmaceutical research and drug development. Current Opinion in Biotechnology. 2019;**55**:51-59

[24] Ogrinc Potocnik N, Porta T, Becker M, Heeren RM, Ellis SR. Use of advantageous, volatile matrices enabled by next-generation high-speed matrix-assisted laser desorption/ionization time-of-flight imaging employing a scanning laser beam. Rapid Communications in Mass Spectrometry. 2015;**29**(23):2195-2203

[25] Castellino S, Groseclose MR, Wagner D. MALDI imaging mass spectrometry: Bridging biology and chemistry in drug development. Bioanalysis. 2011;**3**(21):2427-2441

[26] Second TP, Blethrow JD, Schwartz JC, Merrihew GE, MacCoss MJ, Swaney DL, et al. Dual-pressure linear ion trap mass spectrometer improving the analysis of complex protein mixtures. Analytical Chemistry. 2009;**81**(18):7757-7765

[27] Michalski A, Cox J, Mann M. More than 100,000 detectable peptide species elute in single shotgun proteomics runs but the majority is inaccessible to data-dependent LC-MS/MS. Journal of Proteome Research. 2011;**10**(4):1785-1793

[28] Mann M, Kelleher NL. Precision proteomics: The case for high resolution and high mass accuracy. Proceedings of the National Academy of Sciences of the United States of America. 2008;**105**(47):18132-18138

[29] Beck S, Michalski A, Raether O, Lubeck M, Kaspar S, Goedecke N, et al. The impact II, a very high-resolution quadrupole time-of-flight instrument (QTOF) for deep shotgun proteomics. Molecular & Cellular Proteomics. 2015;**14**(7):2014-2029

[30] Makarov A. Electrostatic axially harmonic orbital trapping: A high-performance technique of mass analysis. Analytical Chemistry. 2000;**72**(6):1156-1162

[31] Hu Q, Noll RJ, Li H, Makarov A, Hardman M, Graham Cooks R. The orbitrap: A new mass spectrometer. Journal of Mass Spectrometry. 2005;**40**(4):430-443

[32] Eliuk S, Makarov A. Evolution of orbitrap mass spectrometry instrumentation. Annual Review of Analytical Chemistry. 2015;**8**:61-80

[33] Makarov A, Denisov E, Kholomeev A, Balschun W, Lange O, Strupat K, et al. Performance evaluation of a hybrid linear ion trap/orbitrap mass spectrometer. Analytical Chemistry. 2006;**78**(7):2113-2120

[34] Thakur SS, Geiger T, Chatterjee B, Bandilla P, Frohlich F, Cox J, et al. Deep and highly sensitive proteome coverage by LC-MS/MS without prefractionation. Molecular & Cellular Proteomics. 2011;**10**(8):M110.003699

[35] Picotti P, Clement-Ziza M, Lam H, Campbell DS, Schmidt A, Deutsch EW, et al. A complete mass-spectrometric map of the yeast proteome applied to quantitative trait analysis. Nature. 2013;**494**(7436):266-270

[36] Arike L, Valgepea K, Peil L, Nahku R, Adamberg K, Vilu R. Comparison and applications of label-free absolute proteome quantification methods on *Escherichia coli*. Journal of Proteomics. 2012;**75**(17):5437-5448

[37] Krisp C, Parker R, Pascovici D, Hayward NK, Wilmott JS, Thompson JF, et al. Proteomic phenotyping of metastatic melanoma reveals putative signatures of MEK inhibitor response and prognosis. British Journal of Cancer. 2018;**119**(6):713-723

[38] Pirmoradian M, Budamgunta H, Chingin K, Zhang B, Astorga-Wells J, Zubarev RA. Rapid and deep human proteome analysis by single-dimension shotgun proteomics. Molecular & Cellular Proteomics. 2013;**12**(11):3330-3338

[39] Michalski A, Damoc E, Hauschild JP, Lange O, Wieghaus A, Makarov A, et al. Mass spectrometry-based proteomics using Q exactive, a high-performance benchtop quadrupole orbitrap mass spectrometer. Molecular & Cellular Proteomics. 2011;**10**(9):M111.011015

[40] Butnaru CM, Chiritoiu MB, Chiritoiu GN, Petrescu SM, Petrescu AJ. Inhibition of N-glycan processing modulates the network of EDEM3 interactors. Biochemical and Biophysical Research Communications. 2017;**486**(4):978-984

[41] Coffill CR, Muller PA, Oh HK, Neo SP, Hogue KA, Cheok CF, et al. Mutant p53 interactome identifies nardilysin as a p53R273H-specific binding partner that promotes invasion. EMBO Reports. 2012;**13**(7):638-644

[42] Bowler EH, Wang Z, Ewing RM. How do oncoprotein mutations rewire protein-protein interaction networks? Expert Review of Proteomics. 2015;**12**(5):449-455

[43] Kawahara R, Meirelles GV, Heberle H, Domingues RR, Granato DC, Yokoo S, et al. Integrative analysis to select cancer candidate biomarkers to targeted validation. Oncotarget. 2015;**6**(41):43635-43652

[44] Schadendorf D, van Akkooi ACJ, Berking C, Griewank KG, Gutzmer R, Hauschild A, et al. Melanoma. Lancet. 2018;**392**(10151):971-984

[45] Tarhini A, Kudchadkar RR. Predictive and on-treatment monitoring biomarkers in advanced melanoma: Moving toward personalized medicine. Cancer Treatment Reviews. 2018;**71**:8-18

[46] Bougnoux AC, Solassol J. The contribution of proteomics to the identification of biomarkers for cutaneous malignant melanoma. Clinical Biochemistry. 2013;**46**(6):518-523

[47] Perego M, Maurer M, Wang JX, Shaffer S, Muller AC, Parapatics K,

et al. A slow-cycling subpopulation of melanoma cells with highly invasive properties. Oncogene. 2018;**37**(3):302-312

[48] Liu H, Ong SE, Badu-Nkansah K, Schindler J, White FM, Hynes RO. CUB-domain-containing protein 1 (CDCP1) activates Src to promote melanoma metastasis. Proceedings of the National Academy of Sciences of the United States of America. 2011;**108**(4):1379-1384

[49] Janostiak R, Rauniyar N, Lam TT, Ou J, Zhu LJ, Green MR, et al. MELK promotes melanoma growth by stimulating the NF-kappaB pathway. Cell Reports. 2017;**21**(10):2829-2841

[50] Gustafsson OJ, Arentz G, Hoffmann P. Proteomic developments in the analysis of formalin-fixed tissue. Biochimica et Biophysica Acta. 2015;**1854**(6):559-580

[51] Fox CH, Johnson FB, Whiting J, Roller PP. Formaldehyde fixation. The Journal of Histochemistry and Cytochemistry. 1985;**33**(8):845-853

[52] Metz B, Kersten GF, Baart GJ, de Jong A, Meiring H, ten Hove J, et al. Identification of formaldehyde-induced modifications in proteins: Reactions with insulin. Bioconjugate Chemistry. 2006;**17**(3):815-822

[53] Metz B, Kersten GF, Hoogerhout P, Brugghe HF, Timmermans HA, de Jong A, et al. Identification of formaldehyde-induced modifications in proteins: Reactions with model peptides. The Journal of Biological Chemistry. 2004;**279**(8):6235-6243

[54] Byrum SD, Larson SK, Avaritt NL, Moreland LE, Mackintosh SG, Cheung WL, et al. Quantitative proteomics identifies activation of hallmark pathways of cancer in patient melanoma. Journal of Proteomics & Bioinformatics. 2013;**6**(3):43-50

[55] Sabel MS, Liu Y, Lubman DM. Proteomics in melanoma biomarker discovery: Great potential, many obstacles. International Journal of Proteomics. 2011;**2011**:181890

[56] Lodish HBA, Zipursky SL, et al. Molecular Cell Biology. New York: W. H. Freeman; 2000

[57] Singh C, Zampronio CG, Creese AJ, Cooper HJ. Higher energy collision dissociation (HCD) product ion-triggered electron transfer dissociation (ETD) mass spectrometry for the analysis of N-linked glycoproteins. Journal of Proteome Research. 2012;**11**(9):4517-4525

[58] Chiritoiu GN, Jandus C, Munteanu CV, Ghenea S, Gannon PO, Romero P, et al. Epitope located N-glycans impair the MHC-I epitope generation and presentation. Electrophoresis. 2016;**37**(11):1448-1460

[59] Lin TE, Bondarenko A, Lesch A, Pick H, Cortes-Salazar F, Girault HH. Monitoring tyrosinase expression in non-metastatic and metastatic melanoma tissues by scanning electrochemical microscopy. Angewandte Chemie International Edition. 2016;**55**(11):3813-3816

[60] Osella-Abate S, Quaglino P, Savoia P, Leporati C, Comessatti A, Bernengo MG. VEGF-165 serum levels and tyrosinase expression in melanoma patients: Correlation with the clinical course. Melanoma Research. 2002;**12**(4):325-334

[61] Adina L, Milac GN. The multiple roles of tyrosinase-related protein-2/L-dopachrome tautomerase in melanoma: Biomarker, therapeutic target, and molecular driver in tumor progression. In: Human Skin Cancers-Pathways, Mechanisms, Targets and Treatments. London, UK: IntechOpen; 2017. pp. 45-83

[62] Popa IL, Milac AL, Sima LE, Alexandru PR, Pastrama F,

Munteanu CV, et al. Cross-talk between dopachrome tautomerase and Caveolin-1 is melanoma cell phenotype-specific and potentially involved in tumor progression. The Journal of Biological Chemistry. 2016;**291**(24):12481-12500

[63] Pol J, Vidova V, Kruppa G, Kobliha V, Novak P, Lemr K, et al. Automated ambient desorption-ionization platform for surface imaging integrated with a commercial Fourier transform ion cyclotron resonance mass spectrometer. Analytical Chemistry. 2009;**81**(20):8479-8487

[64] Wiseman JM, Ifa DR, Song Q, Cooks RG. Tissue imaging at atmospheric pressure using desorption electrospray ionization (DESI) mass spectrometry. Angewandte Chemie. 2006;**45**(43):7188-7192

[65] Wiseman JM, Puolitaival SM, Takats Z, Cooks RG, Caprioli RM. Mass spectrometric profiling of intact biological tissue by using desorption electrospray ionization. Angewandte Chemie. 2005;**44**(43):7094-7097

[66] Manicke NE, Dill AL, Ifa DR, Cooks RG. High-resolution tissue imaging on an orbitrap mass spectrometer by desorption electrospray ionization mass spectrometry. Journal of Mass Spectrometry. 2010;**45**(2):223-226

[67] Rao W, Pan N, Yang Z. Applications of the single-probe: Mass spectrometry imaging and single cell analysis under ambient conditions. Journal of Visualized Experiments. 2016;(112):1-9, e53911

[68] Lazova R, Seeley EH. Proteomic mass spectrometry imaging for skin cancer diagnosis. Dermatologic Clinics. 2017;**35**(4):513-519

[69] Lazova R, Yang Z, El Habr C, Lim Y, Choate KA, Seeley EH, et al. Mass spectrometry imaging can distinguish on a proteomic level between proliferative nodules within a benign congenital nevus and malignant melanoma. The American Journal of Dermatopathology. 2017;**39**(9):689-695

[70] Geyer PE, Holdt LM, Teupser D, Mann M. Revisiting biomarker discovery by plasma proteomics. Molecular Systems Biology. 2017;**13**(9):942

[71] Buchbinder EI, Flaherty KT. Biomarkers in melanoma: Lessons from translational medicine. Trends in Cancer. 2016;**2**(6):305-312

[72] Jakob JA, Bassett RL Jr, Ng CS, Curry JL, Joseph RW, Alvarado GC, et al. NRAS mutation status is an independent prognostic factor in metastatic melanoma. Cancer. 2012;**118**(16):4014-4023

[73] Qendro V, Lundgren DH, Rezaul K, Mahony F, Ferrell N, Bi A, et al. Large-scale proteomic characterization of melanoma expressed proteins reveals nestin and vimentin as biomarkers that can potentially distinguish melanoma subtypes. Journal of Proteome Research. 2014;**13**(11):5031-5040

[74] Weinstein D, Leininger J, Hamby C, Safai B. Diagnostic and prognostic biomarkers in melanoma. The Journal of Clinical and Aesthetic Dermatology. 2014;**7**(6):13-24

[75] Al-Ghoul M, Bruck TB, Lauer-Fields JL, Asirvatham VS, Zapata C, Kerr RG, et al. Comparative proteomic analysis of matched primary and metastatic melanoma cell lines. Journal of Proteome Research. 2008;**7**(9):4107-4118

[76] Welinder C, Pawlowski K, Szasz AM, Yakovleva M, Sugihara Y, Malm J, et al. Correlation of histopathologic characteristics to

protein expression and function in malignant melanoma. PLoS One. 2017;**12**(4):e0176167

[77] Joyce CE, Yanez AG, Mori A, Yoda A, Carroll JS, Novina CD. Differential regulation of the melanoma proteome by eIF4A1 and eIF4E. Cancer Research. 2017;**77**(3):613-622

[78] van den Bent MJ, Smits M, Kros JM, Chang SM. Diffuse infiltrating oligodendroglioma and astrocytoma. Journal of Clinical Oncology. 2017;**35**(21):2394-2401

[79] Ohgaki HBW, Reis R, Hegi M, Kleihues P. Gliosarcoma. 2nd ed. Lyon: IARC Press; 2000

[80] Ohka F, Natsume A, Wakabayashi T. Current trends in targeted therapies for glioblastoma multiforme. Neurology Research International. 2012;**2012**:878425

[81] Thakkar JP, Dolecek TA, Horbinski C, Ostrom QT, Lightner DD, Barnholtz-Sloan JS, et al. Epidemiologic and molecular prognostic review of glioblastoma. Cancer Epidemiology, Biomarkers & Prevention. 2014;**23**(10):1985-1996

[82] Jensen RL. Growth factor-mediated angiogenesis in the malignant progression of glial tumors: A review. Surgical Neurology. 1998;**49**(2):189-195. discussion 96

[83] Heroux MS, Chesnik MA, Halligan BD, Al-Gizawiy M, Connelly JM, Mueller WM, et al. Comprehensive characterization of glioblastoma tumor tissues for biomarker identification using mass spectrometry-based label-free quantitative proteomics. Physiological Genomics. 2014;**46**(13):467-481

[84] Hobbs SK, Shi G, Homer R, Harsh G, Atlas SW, Bednarski MD. Magnetic resonance image-guided proteomics of human glioblastoma multiforme. Journal of Magnetic Resonance Imaging. 2003;**18**(5):530-536

[85] Wang Y, Rudnick PA, Evans EL, Li J, Zhuang Z, Devoe DL, et al. Proteome analysis of microdissected tumor tissue using a capillary isoelectric focusing-based multidimensional separation platform coupled with ESI-tandem MS. Analytical Chemistry. 2005;**77**(20):6549-6556

[86] Schwartz SA, Weil RJ, Thompson RC, Shyr Y, Moore JH, Toms SA, et al. Proteomic-based prognosis of brain tumor patients using direct-tissue matrix-assisted laser desorption ionization mass spectrometry. Cancer Research. 2005;**65**(17):7674-7681

[87] Melchior K, Tholey A, Heisel S, Keller A, Lenhof HP, Meese E, et al. Proteomic study of human glioblastoma multiforme tissue employing complementary two-dimensional liquid chromatography- and mass spectrometry-based approaches. Journal of Proteome Research. 2009;**8**(10):4604-4614

[88] Mirza FA, Shamim MS. Tumour treating fields (TTFs) for recurrent and newly diagnosed glioblastoma multiforme. The Journal of the Pakistan Medical Association. 2018;**68**(10):1543-1545

[89] Khalil AA. Biomarker discovery: A proteomic approach for brain cancer profiling. Cancer Science. 2007;**98**(2):201-213

[90] Wibom C, Pettersson F, Sjostrom M, Henriksson R, Johansson M, Bergenheim AT. Protein expression in experimental malignant glioma varies over time and is altered by radiotherapy treatment. British Journal of Cancer. 2006;**94**(12):1853-1863

[91] Miyauchi E, Furuta T, Ohtsuki S, Tachikawa M, Uchida Y, Sabit H, et al. Identification of blood biomarkers in glioblastoma by SWATH mass spectrometry and quantitative targeted absolute proteomics. PLoS One. 2018;**13**(3):e0193799

[92] Gillet LC, Navarro P, Tate S, Rost H, Selevsek N, Reiter L, et al. Targeted data extraction of the MS/MS spectra generated by data-independent acquisition: A new concept for consistent and accurate proteome analysis. Molecular & Cellular Proteomics. 2012;**11**(6):O111.016717

[93] Schuhmann MU, Zucht HD, Nassimi R, Heine G, Schneekloth CG, Stuerenburg HJ, et al. Peptide screening of cerebrospinal fluid in patients with glioblastoma multiforme. European Journal of Surgical Oncology. 2010;**36**(2):201-207

[94] Benveniste RJ, Germano IM. Correlation of factors predicting intraoperative brain shift with successful resection of malignant brain tumors using image-guided techniques. Surgical Neurology. 2005;**63**(6):542-548. discussion 8-9

[95] Kröger S, Niehoff AC, Jeibmann A, Sperling M, Paulus W, Stummer W, et al. Complementary molecular and elemental mass-spectrometric imaging of human brain tumors resected by fluorescence-guided surgery. Analytical Chemistry. 2018;**90**(20):12253-12260

[96] Idoate MA, Diez Valle R, Echeveste J, Tejada S. Pathological characterization of the glioblastoma border as shown during surgery using 5-aminolevulinic acid-induced fluorescence. Neuropathology. 2011;**31**(6):575-582

[97] Widhalm G, Wolfsberger S, Minchev G, Woehrer A, Krssak M, Czech T, et al. 5-Aminolevulinic acid is a promising marker for detection of anaplastic foci in diffusely infiltrating gliomas with nonsignificant contrast enhancement. Cancer. 2010;**116**(6):1545-1552

[98] Brokinkel B, Kroger S, Senner V, Jeibmann A, Karst U, Stummer W. Visualizing protoporphyrin IX formation in the dura tail of meningiomas by mass spectrometry imaging. Acta Neurochirurgica. 2018;**160**(7):1433-1437

[99] Feigin IH, Gross SW. Sarcoma arising in glioblastoma of the brain. The American Journal of Pathology. 1955;**31**(4):633-653

[100] Vukelić Z, Kalanj-Bognar S, Froesch M, Bindila L, Radic B, Allen M, et al. Human gliosarcoma-associated ganglioside composition is complex and distinctive as evidenced by high-performance mass spectrometric determination and structural characterization. Glycobiology. 2007;**17**(5):504-515

[101] Lutterbach J, Guttenberger R, Pagenstecher A. Gliosarcoma: A clinical study. Radiotherapy and Oncology. 2001;**61**(1):57-64

[102] Kelly KA, Kirkwood JM, Kapp DS. Glioblastoma multiforme: Pathology, natural history and treatment. Cancer Treatment Reviews. 1984;**11**(1):1-26

[103] Prinetti A, Prioni S, Loberto N, Aureli M, Nocco V, Illuzzi G, et al. Aberrant glycosphingolipid expression and membrane organization in tumor cells: Consequences on tumor-host interactions. Advances in Experimental Medicine and Biology. 2011;**705**:643-667

[104] Fredman P, Hedberg K, Brezicka T. Gangliosides as therapeutic targets for cancer. BioDrugs. 2003;**17**(3):155-167

[105] Hakomori S. Glycosylation defining cancer malignancy: New wine in an old bottle. Proceedings of the National Academy of Sciences

of the United States of America. 2002;**99**(16):10231-10233

[106] Hakomori S. Tumor-associated carbohydrate antigens defining tumor malignancy: Basis for development of anti-cancer vaccines. Advances in Experimental Medicine and Biology. 2001;**491**:369-402

[107] Hakomori SI, Murakami WT. Glycolipids of hamster fibroblasts and derived malignant-transformed cell lines. Proceedings of the National Academy of Sciences of the United States of America. 1968;**59**(1):254-261

[108] Malisan F, Testi R. GD3 ganglioside and apoptosis. Biochimica et Biophysica Acta. 2002;**1585**(2-3):179-187

[109] Sung CC, Pearl DK, Coons SW, Scheithauer BW, Johnson PC, Zheng M, et al. Correlation of ganglioside patterns of primary brain tumors with survival. Cancer. 1995;**75**(3):851-859

[110] Fabris D, Rozman M, Sajko T. Vukelic Z. Aberrant ganglioside composition in glioblastoma multiforme and peritumoral tissue: A mass spectrometry characterization. Biochimie. 2017;**137**:56-68

[111] Jennemann R, Rodden A, Bauer BL, Mennel HD, Wiegandt H. Glycosphingolipids of human gliomas. Cancer Research. 1990;**50**(23):7444-7449

[112] Hedberg KM, Mahesparan R, Read TA, Tysnes BB, Thorsen F, Visted T, et al. The glioma-associated gangliosides 3′-isoLM1, GD3 and GM2 show selective area expression in human glioblastoma xenografts in nude rat brains. Neuropathology and Applied Neurobiology. 2001;**27**(6):451-464

[113] Steiner G, Shaw A, Choo-Smith LP, Abuid MH, Schackert G, Sobottka S, et al. Distinguishing and grading human gliomas by IR spectroscopy. Biopolymers. 2003;**72**(6):464-471

[114] Wagener R, Rohn G, Schillinger G, Schroder R, Kobbe B, Ernestus RI. Ganglioside profiles in human gliomas: Quantification by microbore high performance liquid chromatography and correlation to histomorphology and grading. Acta Neurochirurgica. 1999;**141**(12):1339-1345

[115] Zamfir AD. Microfluidics-mass spectrometry of protein-carbohydrate interactions: Applications to the development of therapeutics and biomarker discovery. Methods in Molecular Biology. 2017;**1647**:109-128

[116] Zamfir AD. Neurological analyses: Focus on gangliosides and mass spectrometry. Advances in Experimental Medicine and Biology. 2014;**806**:153-204

[117] Kniep B, Kniep E, Ozkucur N, Barz S, Bachmann M, Malisan F, et al. 9-O-acetyl GD3 protects tumor cells from apoptosis. International Journal of Cancer. 2006;**119**(1):67-73

[118] Zamfir A, Vukelic Z, Bindila L, Peter-Katalinic J, Almeida R, Sterling A, et al. Fully-automated chip-based nanoelectrospray tandem mass spectrometry of gangliosides from human cerebellum. Journal of the American Society for Mass Spectrometry. 2004;**15**(11):1649-1657

[119] Radon TP, Massat NJ, Jones R, Alrawashdeh W, Dumartin L, Ennis D, et al. Identification of a three-biomarker panel in urine for early detection of pancreatic adenocarcinoma. Clinical Cancer Research. 2015;**21**(15):3512-3521

[120] von Neubeck C, Seidlitz A, Kitzler HH, Beuthien-Baumann B, Krause M. Glioblastoma multiforme: Emerging treatments and stratification

markers beyond new drugs. The British Journal of Radiology. 2015;**88**(1053):20150354

[121] Chen HY, Varki A. O-acetylation of GD3: An enigmatic modification regulating apoptosis? The Journal of Experimental Medicine. 2002;**196**(12):1529-1533

[122] Malisan F, Franchi L, Tomassini B, Ventura N, Condo I, Rippo MR, et al. Acetylation suppresses the proapoptotic activity of GD3 ganglioside. The Journal of Experimental Medicine. 2002;**196**(12):1535-1541

[123] Liu X, Zheng W, Wang W, Shen H, Liu L, Lou W, et al. A new panel of pancreatic cancer biomarkers discovered using a mass spectrometry-based pipeline. British Journal of Cancer. 2017;**117**(12):1846-1854

[124] Nigjeh EN, Chen R, Brand RE, Petersen GM, Chari ST, von Haller PD, et al. Quantitative proteomics based on optimized data-independent acquisition in plasma analysis. Journal of Proteome Research. 2017;**16**(2):665-676

[125] Yoneyama T, Ohtsuki S, Honda K, Kobayashi M, Iwasaki M, Uchida Y, et al. Identification of IGFBP2 and IGFBP3 as compensatory biomarkers for CA19-9 in early-stage pancreatic cancer using a combination of antibody-based and LC-MS/MS-based proteomics. PLoS One. 2016;**11**(8):e0161009

[126] Castillo J, Bernard V, San Lucas FA, Allenson K, Capello M, Kim DU, et al. Surfaceome profiling enables isolation of cancer-specific exosomal cargo in liquid biopsies from pancreatic cancer patients. Annals of Oncology. 2018;**29**(1):223-229

[127] Kirana C, Peng L, Miller R, Keating JP, Glenn C, Shi H, et al. Combination of laser microdissection, 2D-DIGE and MALDI-TOF MS to identify protein biomarkers to predict colorectal cancer spread. Clinical Proteomics. 2019;**16**:3

[128] Jimenez L, Yu H, McKenzie A, Franklin JL, Patton JG, Liu Q, et al. Quantitative proteomic analysis of small and large extracellular vesicles (EVs) reveals enrichment of adhesion proteins in small EVs. Journal of Proteome Research. 2019

[129] Kit OI, Vodolazhsky DI, Kutilin DS, Enin YS, Gevorkyan YA, Zolotukhin PV, et al. A proteomics analysis reveals 9 up-regulated proteins associated with altered cell signaling in colon cancer patients. The Protein Journal. 2017;**36**(6):513-522

[130] Lee CH, Im EJ, Moon PG, Baek MC. Discovery of a diagnostic biomarker for colon cancer through proteomic profiling of small extracellular vesicles. BMC Cancer. 2018;**18**(1):1058

[131] Hu HF, Xu WW, Wang Y, Zheng CC, Zhang WX, Li B, et al. Comparative proteomics analysis identifies Cdc42-Cdc42BPA signaling as prognostic biomarker and therapeutic target for colon cancer invasion. Journal of Proteome Research. 2018;**17**(1):265-275

[132] Alvarez-Chaver P, De Chiara L, Martinez-Zorzano VS. Proteomic profiling for colorectal cancer biomarker discovery. Methods in Molecular Biology. 2018;**1765**:241-269

[133] Suwakulsiri W, Rai A, Xu R, Chen M, Greening DW, Simpson RJ. Proteomic profiling reveals key cancer progression modulators in shed microvesicles released from isogenic human primary and metastatic colorectal cancer cell lines, Biochim Biophys Acta Proteins Proteom. 2018. pii: S1570-9639(18)30206-1

Chapter 6

Invasiveness-Related Proteomic Variations and Molecular Network Changes in Human Nonfunctional Pituitary Adenomas

Xianquan Zhan, Xiaohan Zhan and Xiaowei Wang

Abstract

The invasive characteristic of nonfunctional pituitary adenoma (NFPA) is an important clinical problem without a clear molecular mechanism, which severely challenges its treatment strategy. Clarification of the proteomic alterations between invasive and non-invasive NFPAs is the key step for in-depth understanding of its mechanisms and discovering reliably invasive biomarkers. Two-dimensional gel electrophoresis (2DGE)-based comparative proteomics was carried out between four invasive and four non-invasive NFPAs. A total of 64 upregulated protein-spots and 39 downregulated protein-spots were identified among 24 (invasive n = 12; non-invasive n = 12) 2DGE maps (ca. 1200 spots/gel). Mass spectrometry identified 30 upregulated proteins and 27 downregulated proteins between invasive and non-invasive NFPAs. Those 57 differentially expressed proteins are involved in multiple biological functions, including oxidative stress, mitochondrial dysfunction, MAPK signaling alteration, proteolysis abnormality, CDK-C signaling, amyloid processing, and TR/RXR activation. These findings provide important clues to insights into molecular mechanisms of invasive NFPAs and to discovery of effective biomarkers for effective treatment of invasive NFPA patients.

Keywords: invasive nonfunctional pituitary adenoma, two-dimensional gel electrophoresis, mass spectrometry, proteome, comparative proteomics, invasive biomarker

1. Introduction

Invasive pituitary adenoma is a type of pituitary adenoma that locally invades contiguous anatomy structures surrounding pituitary gland [1–6]. In fact, the rate of local invasion is about 40% of pituitary adenoma patients with macroscopic observation, and even up to 80% of pituitary adenoma patients with microscopic observation [1, 7, 8] although most pituitary adenomas are align. Magnetic resonance imaging (MRI) is commonly used method to measure the size of pituitary adenomas, and can classify pituitary adenomas into giant adenomas (>40 mm), macro-plus adenomas (20–30 mm), macroadenomas (10–20 mm), and microadenomas (<10 mm) [5, 7]. Furthermore, based on preoperative MRI and perioperative observation, pituitary adenomas are classified into grade I (enclosed microadenoma, <10 mm), grade II (enclosed macroadenoma, >10 mm), grade III (localized perforation of the sellar floor), and grade IV (diffuse destruction of the sellar floor) [9]. Grades III and IV are commonly looked as invasive pituitary adenomas. Invasiveness is very challenging clinical problem in pituitary adenoma patient, which reasons are that (1) invasiveness suppresses and/or damage surrounding structures because of the limited intracranial cavity and around important structure tissues, and (2) invasiveness causes incomplete removal of pituitary adenoma in neurosurgery to increase risks of complications including recurrence and poor outcome and need adjuvant therapy (radiotherapy or medications) [1]. However, the molecular mechanisms of pituitary adenoma invasiveness remain unclear, although some studies [10] found more vascular evidence in invasive pituitary adenomas compared to non-invasive tumors to indicate the role of angiogenesis [10], and some molecular and genetic changes in invasive pituitary adenomas including downregulation and methylation of CDH13 (H-cadherin) and CDH1 (E-cadherin) [11], loss of death-associated protein kinase and CpG island methylation [12], and loss of heterozygosity at 11q13 (MEN1 locus) and 13q (retinoblastoma gene RB locus) without mutation and overexpression of p53 and without homozygous deletions of p15 or p16 [13]. Multiomics analysis is an effective approach to investigate systematically molecular mechanisms of invasiveness of pituitary adenomas [14–19]. Quantitative transcriptomics analysis [9, 20] identified differentially expressed gene (DEG) profiling (346 DEGs, including 233 upregulated and

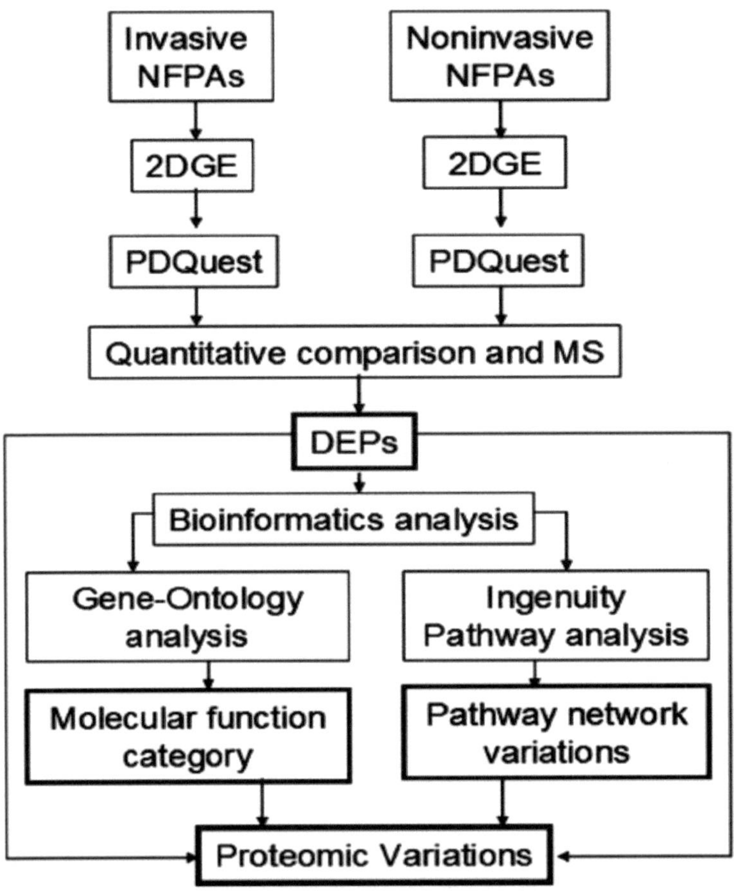

Figure 1.
Experimental flow-chart to comparatively study the proteomes between invasive and non-invasive NFPAs. Reproduced from Zhan et al. [5], with copyright permission from Wiley-VCH, copyright year 2014.

113 downregulated) between invasive and non-invasive NFPAs. However, protein and its proteoforms are the functional performer of each gene, proteome is much more complex than transcriptome, and the coefficient of correlation is very low (about 0.4) in consistence analysis between proteome and transcriptome for the same tissue sample [21, 22]. Therefore, it is necessary to use proteomics for pituitary adenoma invasiveness [23, 24]. A comparative proteomics experiment revealed 30 differentially expressed proteins (DEPs) profiling between invasive and non-invasive pituitary adenoma tissues [25], however, this study did not distinguish the functional and non-functional pituitary adenomas (FPAs, and NFPAs). This chapter focused on the proteomic variations and molecular network changes in invasive relative to noninvasive NFPAs, investigated with two-dimensional gel electrophoresis (2DGE) coupled with mass spectrometry (MS) and pathway network analysis. The findings offer the scientific data to discover protein biomarkers for effective treatment of invasive NFPAs. An experimental flow-chart is shown to study proteomes between invasive and noninvasive NFPAs (**Figure 1**).

2. Materials and methods

2.1 2DGE analysis of pituitary adenoma specimen

The invasive (n = 4) and non-invasive (n = 4) NFPA tissues with pathological diagnosis were used in this study. Each tissue sample was used to individually extract proteins, and the protein content was quantified. Each tissue sample was analyzed with 2DGE for 3–4 times [5, 22]. For each 2DGE analysis, 150 μg proteins were used for isoelectric focusing (IEF) with IPG strips pH 3–10 NL (180 × 3 × 0.5 mm). After IEF, the proteins was reduced and alkalized, and then were separated with the 12% PAGE resolving gel (250 × 215 × 1.0 mm), followed by visualization with modified silver-staining [26]. The PDQuest 2D gel analysis software (version 7.1.0; Bio-Rad) was used to digitize and compare 2DGE gel images between invasive and non-invasive NFPAs. A total of 12 gel images for invasive NFPAs and 12 gel images for non-invasive NFPAs were used in this analysis to determine each DEPs with a 3-fold cutoff values and $p < 0.05$. In addition, four standard proteins, including myoglobin (17 kDa; p. 7.6), carbonic anhydrase (29 kDa; p. 7.0), ovalbumin (45 kDa; p. 5.1), and amyloglucosidase (89/70 kDa; p. 3.8), were applied to measure the observed p*I* and *Mr* on the 2D gel.

2.2 Mass spectrometry analysis of 2DGE-separated proteins

The protein that contains in gel spot was digested in-gel with trypsin, followed by ZipTipC18 purification [5, 26]. For LC-ESI-MS/MS analysis, the purified tryptic peptides were eluted in 6 μl of 85% acetonitrile plus 0.1% TFA, air-dried, and then resuspended in 6 μl of 85% acetonitrile plus 0.1% formic acid. The prepared peptide samples were analyzed by LC-ESI-qTOF mass spectrometer to obtain MS/MS spectrum. For MADI-TOF-MS analysis, the ZipTipC18 peptides were directly eluted on MALDI plate with 2 μl of a-cyano-4-hydroxycinnamic acid solution (seven cycles), and dried, and then were analyzed with Voyager DE STR MALDI-TOF mass spectrometer to obtain peptide mass fingerprint (PMF). The MS/MS data and PMF data were used to search SwissProt database with Mascot software for protein identification.

2.3 Bioinformatics

The software NIHDAVID (version 6.7, http://david.abcc.ncifcrf.gov/summary.jsp) was used to carry out gene-ontology (GO) analysis, including cellular components (CC), molecular functions (MF), and biological processes (BP), and furtherly were categorized into different functional clusters. Ingenuity pathway analysis (IPA) (www.ingenuity.com) [27] was applied to obtain statistically significant signaling pathways with identified DEP data between invasive and non-invasive NFPAs.

3. Results and discussion

3.1 2DE pattern and DEP profile between invasive and noninvasive NFPA proteomes

Each NFPA tissue sample (four invasive NFPAs and four non-invasive NFPAs) was analyzed by 2DGE for 3–5 times to guarantee at least three high-quality gel images. Thus, 24 high-quality 2DGE images (12 gel-images for invasive NFPAs; 12 gel images for non-invasive NFPAs) were obtained. About 1200 spots (an average of 1172 spots for invasive NFPAs and 1213 spots for non-invasive NFPAs) were present in each gel image (**Figure 2**), and most of spots were distributed within pH 4–9 and *Mr* 15–150 kDa [21]. The average between-gel matched percentage was 64% (61–67%) among invasive NFPA gels, and 67% (61–69%) among non-invasive NFPA gels. The positional deviation of the matched-spots was 2.05 ± 0.89 mm in the IEF direction and 1.41 ± 0.65 mm in the SDS-PAGE direction. For each sample, the average correlation coefficient (*r*) of the normalized volumes for

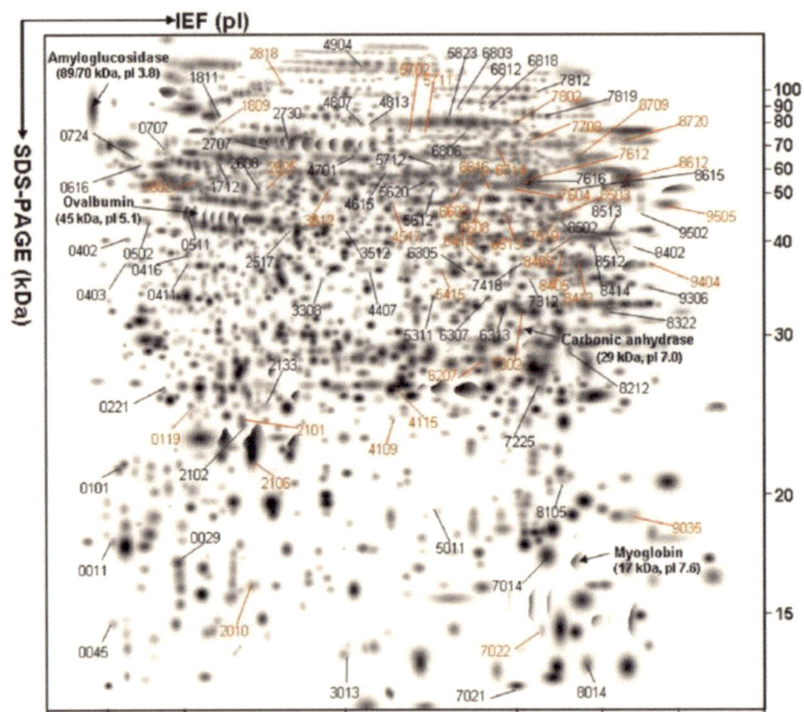

Figure 2.
2DGE map with labeled 4 standard protein markers and 103 spots containing DEPs. IEF was carried out with 18-cm IPGStrip pH 3–10 NL. SDS-PAGE was carried out with 12% polyacrylamide gel. The red means downregulated protein spot in invasive NFPAs relative to noninvasive NFPAs. The black means upregulated spot in invasive NFPAs relative to noninvasive NFPAs. Reproduced from Zhan et al. [5], with copyright permission from Wiley-VCH, copyright year 2014.

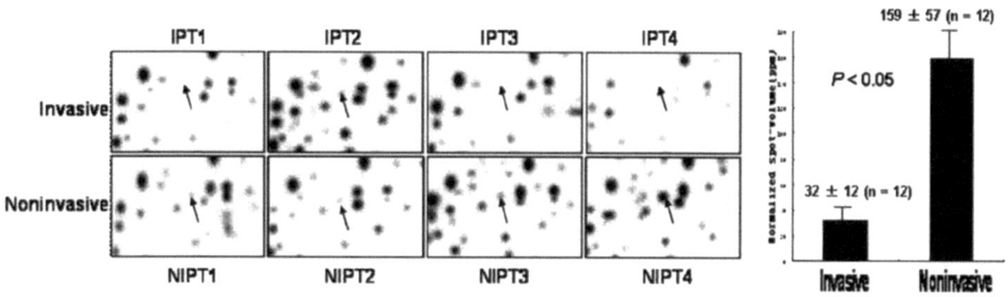

Figure 3.
A presentative differential protein spots between invasive and non-invasive NFPAs (Spot-2010). IPT: Invasive pituitary tumor. NIPT: Non-invasive pituitary tumor. Reproduced from Zhan et al. [5], with copyright permission from Wiley-VCH, copyright year 2014.

between-gel matched-spots was 0.74 (range, 0.59–0.83), with a best-fit line of: y = 0.8685x + 0.0804 (r = 0.87; n = 811). The normalized spot volumes between 12 invasive NFPA gels and 12 non-invasive NFPA gels were compared to determine a differential protein spot with at least 3-fold change and $p < 0.05$. For example, Spot-2010 was identified as differential protein spots downregulated in invasive NFPAs compared to non-invasive NFPAs (**Figure 3**). With the same approach, 103 differential spots were identified, including 64 upregulated and 39 downregulated protein spots in invasive NFPAs relative to non-invasive NFPAs (**Table 1** and **Figure 1**). It clearly demonstrated that the proteome was significantly different between invasive and non-invasive NFPAs.

Furthermore, each DEP in the differential spot was identified with MS [26]. For MALDI-TOF-MS PMF analysis, all interfering masses derived from contaminants including keratins, trypsin, matrix CHCA, and other unknown ones, were removed from MS spectrum of analyzed sample to obtain a corrected mass list for PMF data (**Figure 4**). Those nine masses labeled in **Figure 4B** were used with MASCOT PMF search tool to search Swiss-Prot database, and matched to the corresponding tryptic peptides from 78 kDa glucose-regulated protein (GRP78_HUMAN; P11021) (**Figure 5**), which was the DEP identified in the differential Spot-1809. With the same method, 43 DEPs was identified with PMF analysis (**Figure 1** and **Table 1**). For LC-ESI-MS/MS analysis, the tryptic peptides were separated by LC and then sequenced by MS/MS on the qTOF MS instrument, followed by MASCOT MS/MS data search in the human Swiss-Prot database. For example, six tryptic peptides from Spot-7604 were sequenced and matched to ATP synthase subunit alpha (ATPA_HUMAN; P25705) (**Figure 6**). With the same method, 11 DEPs were identified with MS/MS data (**Figure 1** and **Table 1**). A total of 57 DEPs, including 30 upregulated and 27 downregulated, were identified in invasive compared to non-invasive NFPAs (**Table 1**).

3.2 Functional characteristics of DEPs identified in invasive relative to noninvasive NFPAs

A total of 54 DEPs out of 57 DEPs were eligible for GO analysis to identify the significant BPs, CCs, and MFs, which are further grouped with hierarchical cluster into to functional clusters (**Table 2**). It clearly demonstrated those DEPs participated in multiple biological functions to associate with NFPA invasiveness, including peptidase and proteolysis, nucleotide metabolism, mitochondrial functions and oxidative stress, and protein kinase and cell signaling.

A total of 54 DEPs out of 57 DEPs were accepted for IPA pathway-network analysis to identify significant molecular networks and signaling pathways and

SSP	Swiss-Prot No.	Protein name	Mr (kDa)		pI		Fold
			Exp.	Theor.	Exp.	Theor.	
0011	Q00535	Cyclin-dependent kinase 5	17.56	33.74	4.04	7.57	13.6
0045	P04434	Ig kappa chain V-III region VH (fragment)	14.35	12.86	4.04	5.63	10.2
0029	P00742/ Q8N4Z0	Chain 1: factor X light chain/putative Ras-related protein Rab-42	16.72	54.73/ 11.59	4.90	5.68/ 5.84	3.0
0101	P23297	Protein S100-A1	21.07	10.54	4.12	4.39	18.6
0411*	P04264	Keratin, type II cytoskeletal 1	37.57	66.17	5.05	8.15	5.4
0221	Q5JXM2	Methyltransferase-like protein 24	25.65	41.87	4.71	9.41	4.6
0416	P01040	Cystatin-A	39.04	11.00	5.04	5.38	4.6
0402	Q14314	Fibroleukin	40.94	50.82	4.24	7.08	7.7
0511	P08779/ P04040	Cytokeratin 16/catalase	45.70	51.27/ 59.95	4.89	4.98/ 6.90	16.0
1712	P56817	Beta-secretase 1	61.24	56.36	5.25	5.24	8.5
2608	P78536	Disintegrin and metalloproteinase domain-containing protein 17	52.29	94.56	5.48	5.5	3.5
2133	Q9BYM8	RanBP-type and C3HC4-type zinc finger-containing protein 1	24.67	59.35	5.52	5.47	6.5
2730	Q8N3R9	MAGUK p55 subfamily member 5	68.52	77.53	5.62	5.77	3.2
2707	Q9Y3B9	RRP15-like protein	66.85	31.64	5.53	5.39	3.6
3308	P29466	Caspase-1	35.79	45.81	5.86	5.63	4.0
3013	P07108	Acyl-CoA-binding protein	13.27	10.04	5.98	6.12	3.5
3512	A2VDF0	Fucose mutarotase	42.54	16.93	5.98	5.49	14.4
4407	Q9UIY3	RWD domain-containing protein 2A	37.06	34.21	6.13	6.01	24.3
4701	Q99797	Mitochondrial intermediate peptidase	65.37	81.38	6.04	6.6	12.4
4615	Q96BJ3	Axin interactor, dorsalization- associated protein	56.24	35.17	6.25	6.13	3.4
4807	Q16891	Mitochondrial inner membrane protein	80.13	84.03	6.09	6.08	10.6
7014*	P18988	Hemoglobin beta-2 chain (PANLE)	16.98	15.93	7.33	7.25	3.4
7021*	P06576	ATP synthase subunit beta	12.29	56.56	6.98	5.26	4.2
6313	Q8NA31	Coiled-coil domain-containing protein 79	33.06	84.55	6.92	7.29	3.2
8512	Q8N823	Zinc finger protein 611	41.66	81.39	−1.00	9.16	37.4
8513	Q9Y6N3	Calcium-activated chloride channel regulator family member 3	42.45	30.29	−1.00	8.42	3.2
8212*	Q9P267/ P01834	Methyl-CpG-binding domain protein 5/ Ig kappa chain C region	28.38	159.90/ 11.61	−1.00	9.17/ 5.58	7.8
7616*	Q9P267	Methyl-CpG-binding domain protein 5	55.37	16.12	7.14	9.17	6.1
1602*	A4FU49	SH3 domain-containing protein 21	53.77	70.52	5.13	5.6	−7.4
2010	P60983	Glia maturation factor beta	15.82	16.87	5.45	5.19	−5.0
2106	P01241/ P02792	Chain 1:somatotropin/ferritin light chain	21.87	24.85/ 20.06	5.44	5.29/ 5.51	−5.9
1809	P11021	78 kDa glucose-regulated protein	78.19	72.4	5.24	5.07	−4.6
2101	P01241	Chain 1: somatotropin	23.95	24.85	5.38	5.29	−11.1

SSP	Swiss-Prot No.	Protein name	Mr (kDa)		pI		Fold
			Exp.	Theor.	Exp.	Theor.	
2625	P07332	Tyrosine-protein kinase Fes/Fps	52.98	94.12	5.54	6.27	−7.6
4517	Q8TB05	UBA-like domain-containing protein 1	47.39	19.06	6.26	6.14	−5.6
3612	Q96QD5	DEP domain-containing protein 7	50.55	58.62	5.88	7.62	−4.6
5711	P38405	Guanine nucleotide-binding protein G (olf) subunit alpha	73.36	44.79	6.49	6.23	−4.1
5415	A6NHL2	Tubulin alpha chain-like 3	36.72	50.68	6.56	5.68	−10.7
6207*	P37285	Kinesin light chain 1	27.69	63.74	6.78	5.73	−11.6
5702	P42704	Leucine-rich motif-containing protein, mitochondrial	73.4	159	6.35	5.81	−5.8
6414	Q9UL42	Paraneoplastic antigen Ma2	38.19	41.71	6.8	4.84	−10.3
6513	Q9UPQ3	Arf-GAP with GTPase, ANK repeat and PH domain-containing protein 1	44.44	95.38	6.91	8.18	−17.9
6608	Q7Z3I7/ Q9Y6G9	Zinc finger protein 572/cytoplasmic dynein 1 light intermediate china 1	48.91	63.12/ 56.54	6.75	8.32/ 6.01	−7.3
6603	Q9HD45	Transmembrane 9 superfamily member 3	53.47	68.58	6.68	6.83	−22.5
6616*	P01859	Ig gamma-2 chain C region	52.36	35.9	6.87	7.66	−33.0
7022*	P02080	Hemoglobin beta-C	14.09	15.68	7.27	11.58	−13.8
7604*	P25705	ATP synthase subunit alpha, mitochondrial	53.63	59.83	6.99	9.16	−94.3
7302*	Q96CN7	Isochorismatase domain-containing protein 1	32.25	32.5	6.99	6.96	−3.9
7519	Q99542	Chain 1: matrix metalloproteinase-19	43.84	57.36	7.45	7.22	−7.0
7802	Q96KP1	Exocyst complex component 2	80.76	105.1	6.98	6.46	−16.3
7708	Q02338	D-beta-hydroxybutyrate dehydrogenase, mitochondrial	72.25	38.53	7.11	9.11	−4.9
8503	Q9NPI8	Fanconi anemia group F protein	44.22	42.46	7.53	9.11	−8.0
8405	P17066	Heat shock 70 kDa protein 6	38.49	71.44	7.55	5.81	−25.6
8409	P25101	Endothelin-1 receptor	41.13	49.89	−1.00	8.73	−7.6

*It was identified with LC-ESI-MS/MS, and the others with MALDI-TOF-PMF.
Fold (+) means that it is upregulated in invasive relative to noninvasive NFPAs. Fold (−) means that it is downregulated in invasive relative to noninvasive NFPAs. Exp. pI = −1.00 means that it was out of the pI range of standard markers. Reproduced from Zhan et al. [5], with copyright permission from Wiley-VCH, copyright year 2014.

Table 1.
Differentially expressed proteins between invasive and non-invasive NFPAs identified with 2DGE and mass spectrometry (fold > 3-fold or < −3-fold).

molecular networks. Three molecular networks were identified (**Figure 7**). The hub molecules among those three molecular networks included ATPase, MAPK, ERK, ERK1/2, p38, Jnk, NFkB, AKT, PKA, PKC, EGFR, K-RAS, insulin, UBC, CCND1, IFNG, NFYB, ESR1, CDK5, calmodulin, and S100A1, which are obviously associated with cancer biological systems. About 19 statistically significant canonical pathways were minded from DEPs data (**Figure 8**), including superoxide radical degradation, mitochondrial dysfunction, eNOS signaling, inhibition of matrix metalloprotease, CDK5 signaling, endoplasmic reticulum stress pathway, ketolysis, ketogenesis, TR/RXR activation, amyloid processing, endothelin-1 signaling,

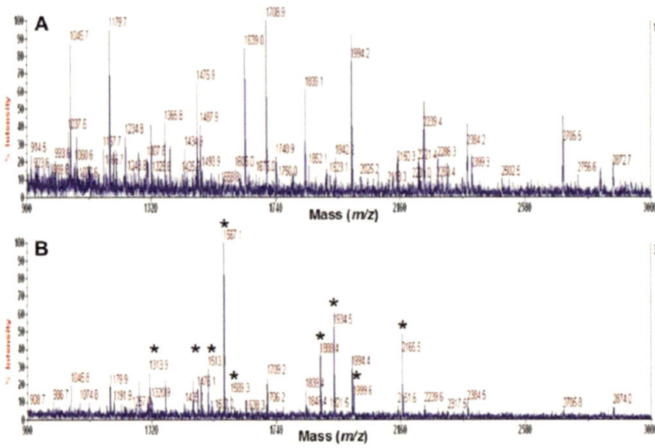

Figure 4.
*All interfering masses from contaminants derived from the margin blank gel on a silver-stained 2D gel map (A) were removed from MALDI-TOF-MS spectrum derived from the proteins in Spot-1809 (B) to obtain a corrected mass list for PMF data that were labeled as the symbol *. Reproduced from Zhan et al. [5], with copyright permission from Wiley-VCH, copyright year 2014.*

A. Summary of Mascot search result

```
User          : Xianquan Zhan
Email         : yjzhan2011@gmail.com
Search title  : Spot-1809
Database      : SwissProt 2013_08 (540732 sequences; 192091492 residues)
Taxonomy      : Homo sapiens (human) (20267 sequences)
Timestamp     : 29 Aug 2013 at 07:07:51 GMT
Top Score     : 119 for GRP78_HUMAN, 78 kDa glucose-regulated protein OS=Homo sapiens GN=HSPA5 PE=1 SV=2
```

Mascot Score Histogram

Protein score is -10*Log(P), where P is the probability that the observed match is a random event.
Protein scores greater than 56 are significant (p<0.05).

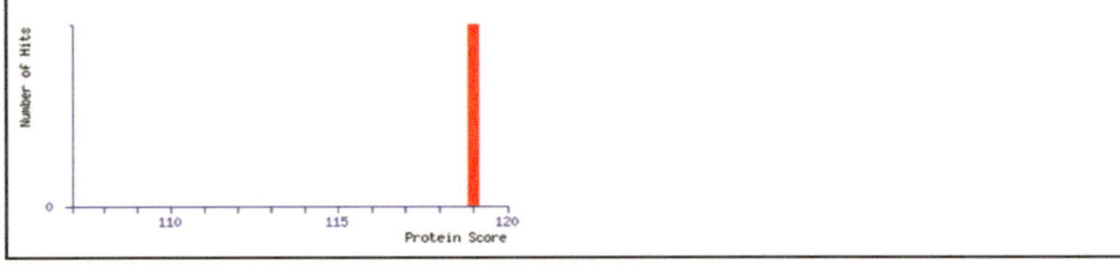

B. Peptides detected by MALDI-TOF-MS

Start-End	Peptide	Observed [M + H]$^+$	Theoretical [M + H]$^+$	Missed cleavage site	Matched
61–74	R.ITPSYVAFTPEGER.L	1567.1409	1567.1336	0	+
165–181	K.VTHAVVTVPAYFNDAQR.Q	1888.4110	1888.4037	0	+
307–324	R.IEIESFYEGEDFSETLTR.A	2165.5073	2165.5000	0	+
325–336	R.AKFEELNMDLFR.S	1513.0857	1513.0784	1	+
327–336	K.FEELNMDLFR.S	1313.9114	1313.9041	0	+
353–367	K.KSDIDEIVLVGGSTR.I	1589.2327	1589.2254	1	+
354–367	K.SDIDEIVLVGGSTR.I	1461.1272	1461.1199	0	+
475–492	K.DNHLLGTFDLTGIPPAPR.G	1934.4548	1934.4475	0	+
493–510	R.GVPQIEVTFEIDVNGILR.V	1999.5542	1999.5469	0	+

C. Matched peptides (Bold) in the amino acid sequence of 78 kDa glucose-regulated protein (GRP78_HUMAN) with a 17% coverage

1 MKLSLVAAML LLLSAARAEE EDKKEDVGTV VGIDLGTTYS CVGVFKNGRV EIIANDQGNR **ITPSYVAFTP** **EGER**LIGDAA KNQLTSNPEN
91 TVFDAKRLIG RTWNDPSVQQ DIKFLPFKVV EKKTKPYIQV DIGGGQTKTF APEEISAMVL TKMKETAEAY LGKK**VTHAVV** **TVPAYFNDAQ**
181 **R**QATKDAGTI AGLNVMRIIN EPTAAAIAYG LDKREGEKNI LVFDLGGGTF DVSLLTIDNG VFEVVATNGD THLGGEDFDQ RVMEHFIKLY
271 KKKTGKDVRK DNRAVQKLRR EVEKAKRALS SQHQAR**IEIE** **SFYEGEDFSE** **TLTRAKFEEL** **NMDLFR**STMK PVQKVLEDSD LK**KSDIDEIV**
361 **LVGGSTR**IPK IQQLVKEFFN GKEPSRGINP DEAVAYGAAV QAGVLSGDQD TGDLVLLDVC PLTLGIETVG GVMTKLIPRN TVVPTKKSQI
451 FSTASDNQPT VTIKVYEGER PLTK**DNHLLG** **TFDLTGIPPA** **PRGVPQIEVT** **FEIDVNGILR** VTAEDKGTGN KNKITITNDQ NRLTPEEIER
541 MVNDAEKFAE EDKKLKERID TRNELESYAY SLKNQIGDKE KLGGKLSSED KETMEKAVEE KIEWLESHQD ADIEDFKAKK KELEEIVQPI
631 ISKLYGSAGP PPTGEEDTAE KDEL

Figure 5.
Mascot search results from PMF data (Spot-1809). Modified from Zhan et al. [5], with copyright permission from Wiley-VCH, copyright year 2014.

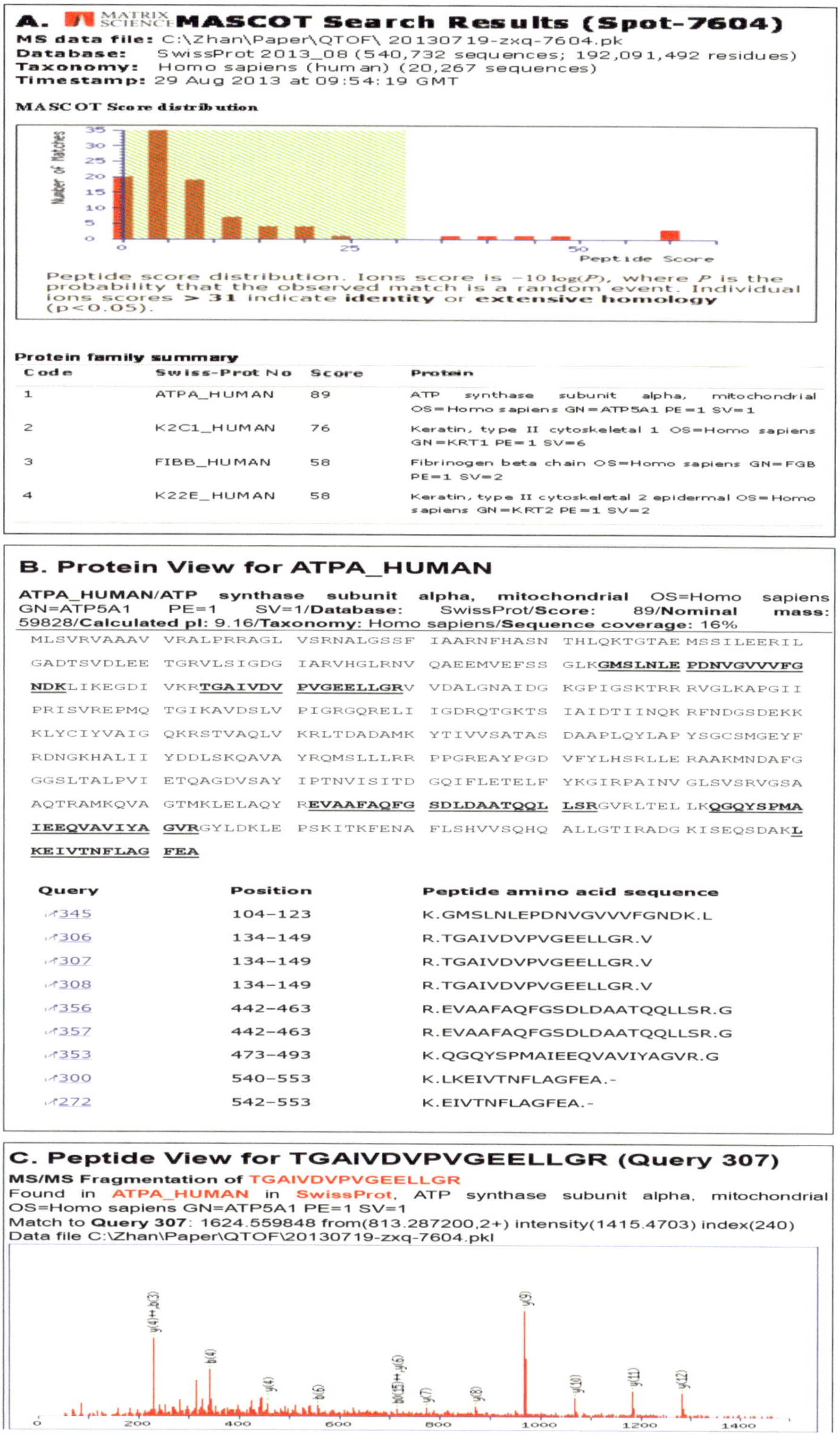

Figure 6.
Mascot search results from a representative LC-ESI-MS/MS data from proteins in Spot-7604. Modified from Zhan et al. [5], with copyright permission from Wiley-VCH, copyright year 2014.

Category	Term	Count	P-value	Proteins (DEPs)
Annotation Cluster 1				
GOTERM_BP_FAT	Regulation of protein kinase cascade	5	5.56E − 03	P29466, Q96BJ3, P00742, P01241, P04040
GOTERM_BP_FAT	Positive regulation of signal transduction	5	1.00E − 02	P29466, P00742, P01241, P04040, P78536
GOTERM_BP_FAT	Positive regulation of protein kinase cascade	4	1.22E − 02	P29466, P00742, P01241, P04040
GOTERM_BP_FAT	Positive regulation of cell communication	5	1.45E-02	P29466, P00742, P01241, P04040, P78536
Annotation Cluster 2				
GOTERM_MF_FAT	Endopeptidase activity	6	3.99E − 03	P29466, P00742, Q99542, Q99797, P56817, P78536
GOTERM_MF_FAT	Peptidase activity, acting on L-amino acid peptides	6	1.89E − 02	P29466, P00742, Q99542, Q99797, P56817, P78536
GOTERM_MF_FAT	Peptidase activity	6	2.25E − 02	P29466, P00742, Q99542, Q99797, P56817, P78536
GOTERM_BP_FAT	Proteolysis	8	2.92E − 02	P29466, P00742, Q99542, Q9BYM8, Q99797, P04264, P56817, P78536
GOTERM_MF_FAT	Metalloendopeptidase activity	3	3.53E − 02	Q99542, Q99797, P78536
Annotation Cluster 3				
GOTERM_CC_FAT	Mitochondrial lumen	5	3.29E − 03	Q02338, P06576, P42704, P25705, Q99797
GOTERM_CC_FAT	Mitochondrial matrix	5	3.29E − 03	Q02338, P06576, P42704, P25705, Q99797
GOTERM_CC_FAT	Mitochondrial part	7	5.08E − 03	Q02338, P06576, P42704, P25705, Q99797, P04040, Q16891
GOTERM_CC_FAT	Organelle envelope	7	6.20E − 03	Q02338, P06576, P42704, P25705, P04040, P25101, Q16891
GOTERM_CC_FAT	Envelope	7	6.29E − 03	Q02338, P06576, P42704, P25705, P04040, P25101, Q16891
GOTERM_CC_FAT	Organelle inner membrane	5	1.20E − 02	Q02338, P06576, P42704, P25705, Q16891
GOTERM_CC_FAT	Mitochondrial envelope	5	2.67E − 02	Q02338, P06576, P25705, P04040, Q16891
GOTERM_CC_FAT	Organelle membrane	8	2.68E − 02	Q02338, P06576, P42704, P25705, P11021, P04040, P25101, Q16891
GOTERM_CC_FAT	Mitochondrial membrane part	3	4.58E − 02	P06576, P25705, Q16891
GOTERM_CC_FAT	Mitochondrial inner membrane	4	5.06E − 02	Q02338, P06576, P25705, Q16891
Annotation Cluster 4				
GOTERM_BP_FAT	Response to organic substance	7	1.63E − 02	P29466, Q00535, P01241, P38405, P25101, P17066, P78536
Annotation Cluster 5				

Category	Term	Count	P-value	Proteins (DEPs)
GOTERM_BP_FAT	Proteolysis	8	2.92E − 02	P29466, P00742, Q99542, Q9BYM8, Q99797, P04264, P56817, P78536
GOTERM_BP_FAT	Protein processing	3	4.13E − 02	P29466, Q99797, P04264
GOTERM_BP_FAT	Protein maturation	3	4.81E − 02	P29466, Q99797, P04264
Annotation Cluster 6				
GOTERM_BP_FAT	Response to alkaloid	3	1.05E − 02	Q00535, P38405, P25101
GOTERM_BP_FAT	Response to organic substance	7	1.63E − 02	P29466, Q00535, P01241, P38405, P25101, P17066, P78536
GOTERM_BP_FAT	Positive regulation of molecular function	6	2.56E − 02	Q00535, P01241, P38405, P04040, P25101, P78536
GOTERM_BP_FAT	Positive regulation of protein kinase activity	4	2.61E − 02	Q00535, P01241, P25101, P78536
GOTERM_BP_FAT	Positive regulation of protein amino acid phosphorylation	3	2.71E − 02	P01241, P25101, P78536
GOTERM_BP_FAT	Positive regulation of kinase activity	4	2.86E − 02	Q00535, P01241, P25101, P78536
GOTERM_BP_FAT	Positive regulation of transferase activity	4	3.15E − 02	Q00535, P01241, P25101, P78536
GOTERM_BP_FAT	Positive regulation of phosphorylation	3	3.18E − 02	P01241, P25101, P78536
GOTERM_BP_FAT	Positive regulation of phosphate metabolic process	3	3.36E − 02	P01241, P25101, P78536
GOTERM_BP_FAT	Positive regulation of phosphorus metabolic process	3	3.36E − 02	P01241, P25101, P78536
GOTERM_BP_FAT	Response to organic cyclic substance	3	4.74E − 02	Q00535, P38405, P25101
Annotation Cluster 7				
GOTERM_MF_FAT	Purine ribonucleotide binding	11	2.91E − 02	Q9Y6G9, P06576, P07332, Q8N4Z0, Q00535, P25705, Q9UPQ3, P11021, P38405, P17066, A6NHL2
GOTERM_MF_FAT	Ribonucleotide binding	11	2.91E − 02	Q9Y6G9, P06576, P07332, Q8N4Z0, Q00535, P25705, Q9UPQ3, P11021, P38405, P17066, A6NHL2
GOTERM_MF_FAT	Purine nucleotide binding	11	3.80E − 02	Q9Y6G9, P06576, P07332, Q8N4Z0, Q00535, P25705, Q9UPQ3, P11021, P38405, P17066, A6NHL2
GOTERM_MF_FAT	Nucleotide binding	12	4.37E − 02	Q9Y6G9, P06576, P07332, Q8N4Z0, Q00535, P25705, Q9UPQ3, P11021, P38405, P04040, P17066, A6NHL2
Annotation Cluster 8				

Category	Term	Count	P-value	Proteins (DEPs)
GOTERM_BP_FAT	Proteolysis	8	2.92E − 02	P29466, P00742, Q99542, Q9BYM8, Q99797, P04264, P56817, P78536
Annotation Cluster 9				
GOTERM_MF_FAT	Calcium ion binding	7	4.37E − 02	P06576, P00742, Q9Y6N3, Q99542, P23297, P11021, Q99797
Annotation Cluster 10				
GOTERM_CC_FAT	Cell surface	5	1.45E − 02	P06576, P00742, P11021, P56817, P78536

Modified from Zhan et al. [5], with copyright permission from Wiley-VCH, copyright year 2014.

Table 2.
The functional categories of 54 DEPs identified by GO analysis.

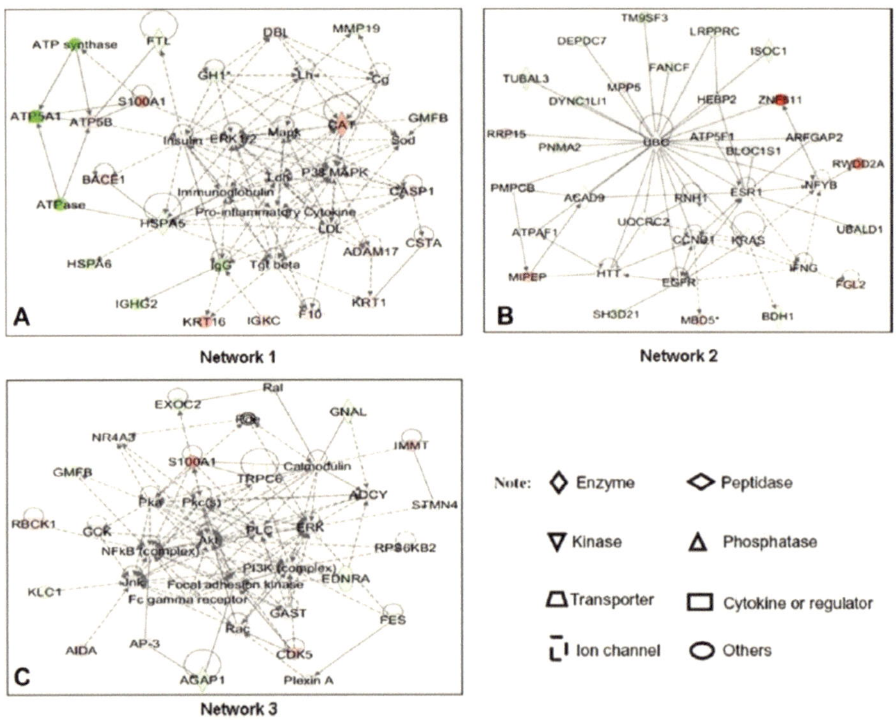

Figure 7.
Significant molecular networks changed in invasive NFPAs. (A) Network 1 functioned in inflammatory disease and inflammatory response. (B) Network 2 functioned in tumor morphology, cancer, cell-to-cell signaling, and interaction. (C) Network 3 functioned in tissue morphology, nervous system development and function, and organismal development. A black solid edge means a direct relationship. A black unsolid edge means an indirect relationship. A red node means upregulated proteins. A green node means downregulated proteins. Reproduced from Zhan et al. [5], with copyright permission from Wiley-VCH, copyright year 2014.

semaphoring signaling in neurons, axonal guidance signaling, neuregulin signaling, and primary immunodeficiency signaling [5]. Also, 10 significant toxicological events were identified with those DEP data, including mitochondrial dysfunction, decreased permeability transition/transmembrane potential/depolarization of mitochondria and mitochondrial membrane, anti-oxidative response panel, and TR/RXR activation. Our previous studies also revealed that MAPK-signaling

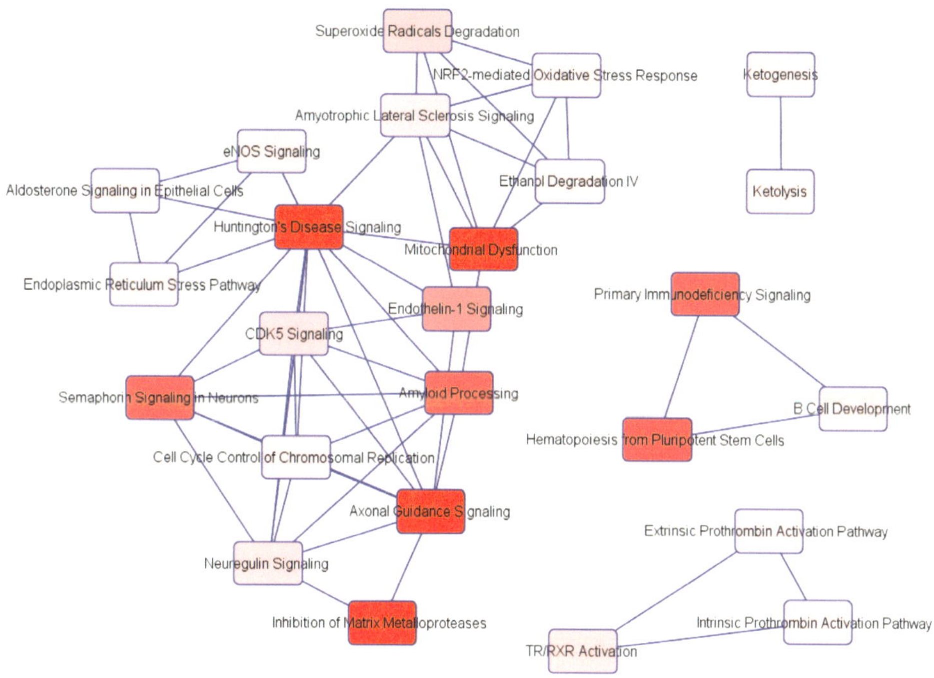

Figure 8.
Statistically significant canonical pathways to involve DEPs in invasive NFPAs. Modified from Zhan et al. [5], with copyright permission from Wiley-VCH, copyright year 2014.

abnormality, oxidative stress, mitochondrial dysfunction, and TR/RXR activation are significantly associated with NFPAs and invasive NFPAs [27], and the changed molecule-pattern in each pathway-system was different between NFPA and invasive NFPA, which might contribute to the pathological processes of invasive NFPAs. Furthermore, ketogenesis and ketolysis, proteolysis abnormality, amyloid processing, and CDK5 signaling abnormality were also obviously related to invasive NAPFs. Therefore MAPK-signaling abnormality, mitochondrial dysfunction, TR/RXR activation, oxidative stress, proteolysis abnormality, CDK5 signaling abnormality, ketogenesis and ketolysis, and amyloid processing were significantly associated with invasive characteristics of invasive NFPAs, and pathway-network-based molecule patterns benefit to identify reliable biomarkers for invasive NFPAs.

4. Conclusions

Invasiveness is serious clinical problem in human pituitary adenomas. It is necessary to clarify its molecular mechanisms and discover effective biomarkers to guide management of invasive NFPAs. This 2DGE-based comparative proteomics and bioinformatics successfully identified proteomic variation profiling and pathway-network changes in human invasive NFPAs compared to noninvasive NFPAs, found 103 differential protein spots (64 upregulated and 39 downregulated) in invasive versus noninvasive NFPA 2DE maps, and identified 57 DEPs (30 upregulated and 27 downregulated), which are significantly involved in pathogenetic process of invasive NFPAs, with altered pathway networks including MAPK-signaling abnormality, oxidative stress, mitochondrial dysfunction, ketogenesis and ketolysis, CDK5 signaling abnormality, TR/RXR activation, proteolysis abnormality, and amyloid processing. Moreover, some important hub-molecules were identified to associate with cancer biological processes, including ATPase,

MAPK, ERK, ERK1/2, p38, Jnk, NFkB, AKT, PKA, PKC, EGFR, K-RAS, insulin, UBC, CCND1, IFNG, NFYB, ESR1, CDK5, calmodulin, and S100A1. Those DEPs, changed pathway networks, and hub-molecules provided new insights into molecular mechanisms of NFPA invasiveness, and important resource for discovery of effective biomarkers to guide the management of invasive NFPAs.

Acknowledgements

The authors acknowledge the financial supports from the Hunan Provincial "Hundred Talent Plan" program (to X.Z.), the Xiangya Hospital Funds for Talent Introduction (to X.Z.), the Hunan Provincial Natural Science Foundation of China (Grant No. 14JJ7008 to X.Z.), China "863" Plan Project (Grant No. 2014AA020610-1 to X.Z.), and the National Natural Science Foundation of China (Grant No. 81572278 and 81272798 to X.Z.). The scientific contributions of Dr. Dominic M. Desiderio from University of Tennessee Health Science Center were also acknowledged.

Conflict of interest

We declare that we have no financial and personal relationships with other people or organizations.

Author's contributions

X.Z. conceived the concept, designed the book chapter, and wrote and critically revised the book chapter, coordinated and was responsible for the correspondence work and financial support. X.H.Z and X.W participated in experiments. X.H.Z edited the English language. All authors approved the final manuscript.

Acronyms and abbreviations

BP	biological processes
CC	cellular components
DEP	differentially expressed protein
ESI	electrospray ionization
FPA	functional pituitary adenomas
IEF	isoelectric focusing
IPA	ingenuity pathway analysis
IPG	immobilized pH gradient
LC	liquid chromatography
MALDI	matrix-assisted laser desorption/ionization
MF	molecular functions
Mr	relative mass
MRI	magnetic resonance imaging
MS	mass spectrometry
MS/MS	tandem mass spectrometry
NFPA	nonfunctional pituitary adenoma
pI	isoelectric point
PMF	peptide mass fingerprint
SDS-PAGE	sodium dodecyl sulfate-polyacrylamide gel electrophoresis
TOF	time-of-flight
2DGE	two-dimensional gel electrophoresis

Author details

Xianquan Zhan[1,2]*, Xiaohan Zhan[1,2] and Xiaowei Wang[1,2]

1 Key Laboratory of Cancer Proteomics of Chinese Ministry of Health, Xiangya Hospital, Central South University, Changsha, P.R. China

2 State Local Joint Engineering Laboratory for Anticancer Drugs, Xiangya Hospital, Central South University, Changsha, P.R. China

*Address all correspondence to: yjzhan2011@gmail.com

References

[1] Hussaini IM, Trotter C, Zhao Y, Abdel-Fattah R, Amos S, Xiao A, et al. Matrix metalloproteinase-9 is differentially expressed in nonfunctioning invasive and noninvasive pituitary adenomas and increases invasion in human pituitary adenoma cell line. The American Journal of Pathology. 2007;**170**:356-365. DOI: 10.2353/ajpath.2007.060736

[2] Martins AN, Hayes GJ, Kempe LG. Invasive pituitary adenomas. Journal of Neurosurgery. 1965;**22**:268-276. DOI: 10.3171/jns.1965.22.3.0268

[3] Zhang X, Fei Z, Zhang W, Zhang JN, Liu WP, Fu LA, et al. Endoscopic endonasal transsphenoidal surgery for invasive pituitary adenoma. Journal of Clinical Neuroscience. 2008;**15**:241-245. DOI: 10.1016/j.jocn.2007.03.008

[4] Hashimoto N, Handa H, Yamashita J, Yamagami T. Long-term follow-up of large or invasive pituitary adenomas. Surgical Neurology. 1986;**25**:49-54. DOI: 10.1016/0090-3019(86)90114-X

[5] Zhan X, Desiderio DM, Wang X, Zhan X, Guo T, Li M, et al. Identification of the proteomic variations of invasive relative to non-invasive non-functional pituitary adenomas. Electrophoresis. 2014;**35**:2184-2194. DOI: 10.1002/elps.201300590

[6] Zhan X, Wang X, Cheng T. Human pituitary adenoma proteomics: New progresses and perspectives. Frontiers in Endocrinology. 2016;**7**:54. DOI: 10.3389/fendo.2016.00054

[7] Meij BP, Lopes MB, Ellegala DB, Alden TD, Laws ER Jr. The long-term significance of microscopic dural invasion in 354 patients with pituitary adenomas treated with transsphenoidal surgery. Journal of Neurosurgery. 2002; **96**:195-208. DOI: 10.3171/jns.2002.96.2.0195

[8] Selman WR, Laws ER Jr, Scheithauer BW, Carpenter SM. The occurrence of dural invasion in pituitary adenomas. Journal of Neurosurgery. 1986;**64**:402-407. DOI: 10.3171/jns.1986.64.3.0402

[9] Galland F, Lacroix L, Saulnier P, Dessen P, Meduri G, Bernier M, et al. Differential gene expression profiles of invasive and non-invasive non-functioning pituitary adenomas based on microarray analysis. Endocrine-Related Cancer. 2010;**17**:361-371. DOI: 10.1677/ERC-10-0018

[10] Turner HE, Nagy Z, Gatter KC, Esiri MM, Harris AL, Wass JA. Angiogenesis in pituitary adenomas—Relationship to endocrine function, treatment and outcome. Journal of Endocrinology. 2000;**165**:475-481

[11] Qian ZR, Sano T, Yoshimoto K, Asa SL, Yamada S, Mizusawa N, et al. Tumor-specific downregulation and methylation of the CDH13 (H-cadherin) and CDH1 (E-cadherin) genes correlate with aggressiveness of human pituitary adenomas. Modern Pathology. 2007;**20**:1269-1277. DOI: 10.1038/modpathol.3800965

[12] Simpson DJ, Clayton RN, Farrell WE. Preferential loss of death associated protein kinase expression in invasive pituitary tumours is associated with either CpG island methylation or homozygous deletion. Oncogene. 2002;**21**:1217-1224. DOI: 10.1038/sj.onc.1205195

[13] Nam DH, Song SY, Park K, Kim MH, Suh YL, Lee JI, et al. Clinical significance of molecular genetic changes in sporadic invasive pituitary adenomas. Experimental & Molecular Medicine. 2001;**33**:111-116. DOI: 10.1038/emm.2001.20

[14] Farrell WE. Pituitary tumours: Findings from whole genome analyses.

Endocrine-Related Cancer. 2006;**13**: 707-716. DOI: 10.1677/erc.1.01131

[15] Zhan X, Desiderio DM. Editorial: Systems biological aspects of pituitary Tumors. Frontiers in Endocrinology. 2016;7:86. DOI: 10.3389/fendo.2016.00086

[16] Zhan X, Long Y. Exploration of molecular network variations in different subtypes of human non-functional pituitary adenomas. Frontiers in Endocrinology. 2016;7:13. DOI: 10.3389/fendo.2016.00013

[17] Hu R, Wang X, Zhan X. Multi-parameter systematic strategies for predictive, preventive and personalised medicine in cancer. The EPMA Journal. 2013;**4**:2. DOI: 10.1186/1878-5085-4-2

[18] Cheng T, Zhan X. Pattern recognition for predictive, preventive, and personalized medicine in cancer. The EPMA Journal. 2017;**8**:51-60. DOI: 10.1007/s13167-017-0083-9

[19] Lu M, Zhan X. The crucial role of multiomic approach in cancer research and clinically relevant outcomes. The EPMA Journal. 2018;**9**:77-102. DOI: 10.1007/s13167-018-0128-8

[20] Wierinckx A, Auger C, Devauchelle P, Reynaud A, Chevallier P, Jan M, et al. A diagnostic marker set for invasion, proliferation, and aggressiveness of prolactin pituitary tumors. Endocrine-Related Cancer. 2007;**14**:887-900. DOI: 10.1677/ERC-07-0062

[21] Zhan X, Desiderio DM. Comparative proteomics analysis of human pituitary adenomas: Current status and future perspectives. Mass Spectrometry Reviews. 2005;**24**: 783-813. DOI: 10.1002/mas.20039

[22] Moreno CS, Evans CO, Zhan X, Okor M, Desiderio DM, Oyesiku NM. Novel molecular signaling and classification of human clinically nonfunctional pituitary adenomas identified by gene expression profiling and proteomic analyses. Cancer Research. 2005;**65**:10214-10222. DOI: 10.1158/0008-5472.CAN-05-0884

[23] Zhan X, Huang Y, Long Y. Two-dimensional gel electrophoresis coupled with mass spectrometry methods for an analysis of human pituitary adenoma tissue proteome. Journal of Visualized Experiments. 2018;**134**:1. DOI: 10.3791/56739

[24] Zhan X, Desiderio DM. The use of variations in proteomes to predict, prevent, and personalize treatment for clinically nonfunctional pituitary adenomas. The EPMA Journal. 2010;**1**: 439-459. DOI: 10.1007/s13167-010-0028-z

[25] Liu Z, Liu Y, Fang W, Chen W, Li C, Xiao Z. Establishment of differential expression profiles from invasive and non-invasive pituitary adenomas. Journal of central south university (medical sciences). 2009;**34**:569-575. DOI: 1672-7347(2009)07-0569-07

[26] Zhan X, Desiderio DM. A reference map of a human pituitary adenoma proteome. Proteomics. 2003;**3**:699-713. DOI: 10.1002/pmic.200300408

[27] Zhan X, Desiderio DM. Signaling pathway networks mined from human pituitary adenoma proteomics data. BMC Medical Genomics. 2010;**3**:13. DOI: 10.1186/1755-8794-3-13